A PRACTICAL APPROACH TO USING LEARNING STYLES IN MATH INSTRUCTION

A PRACTICAL APPROACH TO USING LEARNING STYLES IN MATH INSTRUCTION

By

RUBY BOSTICK MIDKIFF, Ed.D.

Assistant Professor of Education
Arkansas State University
State University, Arkansas

REBECCA DAVIS THOMASSON, Ed.D.

Educational Consultant
National Reading Styles Institute

CHARLES C THOMAS • PUBLISHER
Springfield • Illinois • U.S.A.

Published and Distributed Throughout the World by

CHARLES C THOMAS • PUBLISHER
2600 South First Street
Springfield, Illinois 62794-9265

© *1993 by* CHARLES C THOMAS • PUBLISHER

ISBN 0-398-05888-1

Library of Congress Catalog Card Number: 93-5643

With THOMAS BOOKS *careful attention is given to all details of manufacturing
and design. It is the Publisher's desire to present books that are satisfactory as to their
physical qualities and artistic possibilities and appropriate for their particular use.*
THOMAS BOOKS *will be true to those laws of quality that assure a good name
and good will.*

*Printed in the United States of America
SC-R-3*

Library of Congress Cataloging-in-Publication Data

Midkiff, Ruby Bostick, 1954–
    A practical approach to using learning styles in math instruction
/ by Ruby Bostick Midkiff, Rebecca Davis Thomasson.
      p.   cm.
    Includes bibliographical references and index.
    ISBN 0-398-05888-1
    1. Mathematics—Study and teaching.   2. Cognitive styles.
3. Learning, Psychology of.   I. Thomasson, Rebecca Davis.
II. Title.
QA11.M4965   1994
370.15'651—dc20   93-5643
                                                          CIP

To my parents, W. L. Roach and Doris Young Roach,
my husband, Louis Midkiff, and my son, Martin John Bostick,
with love and pride for their continued support and cooperation.
Ruby Bostick Midkiff

To my husband, Bob Thomasson; my children, Ann Marie Crawford and
Davis Kelly; and my parents, Opheila Barbee Davis and the late Van Davis
because they have always expressed sincere support
for my educational endeavors.
Rebecca Davis Thomasson

and

To all of the teachers who strive to make a difference in lives of children
through the use of learning styles.

ACKNOWLEDGMENTS

Without the love, support, patience, and understanding of our families, this work would not have been completed. Our thanks also go to the many people who made this project a reality, especially the teachers from Blytheville, Arkansas, who participated in the Techniques for Teaching Students Through Their Learning Styles Seminar during the spring of 1990. Their willingness to try new and different techniques and their permission to include some of their activities are greatly appreciated.

CONTENTS

A PRACTICAL APPROACH TO USING LEARNING STYLES IN MATH INSTRUCTION

Chapter One

INTRODUCTION

The need for improvement in mathematics has received increased attention during recent years. If students are to be prepared to meet the needs of the 21st century, mathematics instruction must shift from mastery of abstract facts to a more thorough understanding of mathematical concepts and skills. Decision making, problem solving, and use of technology will be prerequisites for success in almost any career. Furthermore, much attention has been given to the use of learning styles in the general curriculum and in teaching students to read. However, the use of learning styles based instruction in the mathematics classroom has received limited attention by educational authorities. Therefore, the purpose of this book is to address the improvement of mathematics instruction through the use of learning styles based instruction.

The authors of this text advocate implementation of the National Council of Teachers of Mathematics (NCTM) *Curriculum and Evaluation Standards For School Mathematics* (1989) and recommend that mathematics teachers of all levels of instruction read this document and implement use of the described techniques and strategies. Furthermore, consideration of learning styles based instruction as the *NCTM Standards* are studied and implemented in mathematics classrooms is recommended to increase student achievement in mathematics, as well as improve attitudes towards mathematics and students' self-esteem regarding mathematics.

Chapter Two of the text documents the need for improvement in mathematics and addresses how both the curriculum and instruction must change to prepare our students to live successful lives in the 21st century. Chapter Three provides a comprehensive overview of learning styles and is the most substantial chapter of the book since understanding learning styles is necessary for implementation of the strategies described throughout this writing. Effective use of manipulatives is discussed in Chapter Four and a sample lesson plan which can be used as a model for future use has been included. The use of spatial reasoning as a way to reduce gender differences in mathematics achievement is addressed in Chapter Five. For easy reference by the reader, references for additional activities to foster visual thinking and spatial reasoning skills have been included in this chapter in addition to their inclusion in the references section of the book. In order to show how a variety of learning styles can be accommodated while providing appropriate drill-and-practice for students, activities which aid retention of mathematical concepts and skills through a variety of learning styles are presented in Chapter Six. The need for changes in assessment in mathematics and ways to implement authentic assessment through the use of portfolios are presented in Chapter Seven. The final chapter summarizes the key ideas presented in detail throughout the book. References for all of the chapters are included in the last segment of this book. All of the chapters have been divided by subheadings including a conclusion which summarizes the key concepts of each section. These subheadings are listed in the table of contents for easy reference by the reader. Specific strategies and activities have been indented throughout the book so that they can be easily identified by the reader. Further explanations of these strategies and activities often follow the indentations in regular format.

The goals of this book, therefore, are to give the reader an understanding of learning styles based instruction in mathematics, of effective use of manipulatives in teaching various concepts at all grade levels, of ways to develop spatial reasoning skills in students, of different activities which accommodate a variety of learning styles, and of authentic assessment in mathematics. The authors contend that use of learning styles based instruction is a powerful strategy which teachers can and should utilize. When the methodology discussed in this book is utilized, teaching will be more effective, less remediation will be necessary, and the overall mathematics curriculum will be enhanced.

The authors, as experienced teachers, have used the suggested activities and approaches described within and can attest to their utility, versatility, and overall effectiveness. Readers are encouraged to modify the suggestions as appropriate and to be creative and flexible in their conceptualization and adaptations of this book's content. Teaching and learning are both complex processes that will become increasingly better understood over time and which continue to require judgement, energy, motivation, and hard work.

Chapter Two

IMPROVING MATHEMATICS INSTRUCTION

NEED FOR INCREASED ACHIEVEMENT IN MATHEMATICS

In recent years, basic skills tests, including a mathematics component, have been implemented in many states and, as a result, a great deal of emphasis has been placed on accountability in teaching. In the same realm, teachers are showing increasing concern about student academic achievement in mathematics. There is a growing number of students entering colleges and universities in need of remedial mathematics. In fact, *Everybody Counts: A Report to the Nation on the Future of Mathematics Education* (National Research Council, 1989) discusses the serious situation that our country faces and the need for high-quality instruction in mathematics. This report also states that the first high school graduates of the 21st century "will leave school without sufficient preparation in mathematics to cope with either on-the-job demands for problem solving or college expectations for mathematical literacy" (p. 2). The National Education Goals for the year 2000 (1989) also state, "Goal 4 Science and Mathematics. U.S. Students will be first in the world in science and mathematics achievement." Furthermore, the *National Council of Teachers of Mathematics Curriculum and Evaluation Standards for School Mathematics* (1989) were created "as one means to help improve the quality of school mathematics" (p. v).

Several approaches can be used to remedy the problem of low student performance in mathematics. All students must receive mathematics instruction and preparation appropriate for the 21st century. Teachers must understand and emphasize that knowledge about and high academic performance in mathematics is the best way "up" for students from disadvantaged backgrounds (National Research Council, 1989). The employees of the next century must be "problem solvers, decision makers, adept negotiators, and thinkers who are at home with open-endedness, flexibility, and resourcefulness" (Caine & Caine, 1991, p. 14). Furthermore, problem-solving skills, use of the calculator and manipula-

5

tives in instruction, active engagement of students in the learning process, and mathematics based on real life experiences are discussed in detail in the NCTM Standards (1989). In a report prepared by the Arkansas Governor's Office (1992) this need in mathematics instruction is also addressed: "Instead of students learning information by rote, they must become better problem solvers, learning to apply the information to real world situations" (p. 6).

Recent reports concerning mathematics education have also revealed that an inability to do as well in mathematics as a student does in other subjects is not only accepted by some parents, but expected. They never did well in math and do not expect their children to be mathematicians. Many people simply believe that the majority of today's students cannot do math. The result of these lower expectations is that being a poor math student is socially acceptable. Perhaps worse is the peer pressure that makes superior mathematics performance unacceptable and, furthermore, these negative expectations are strongest among minorities and women (see Chapter 5). In fact, most of the public do not have fond memories of high school or college math classes and, perhaps, lower level math classes. They remember that they took the class(es) they had to take to graduate and, gratefully, never took another math class. Therefore, many parents may accept and even expect their children to perform poorly in mathematics (National Research Council, 1989).

Motivating students to learn and making learning meaningful are two factors that are the key to any child's success in mathematics. However, despite a wealth of well-conducted research in the areas of math instruction, including the NCTM Standards (1989), some teachers continue to function in the "dark ages" in terms of the teaching-learning process. Many teachers continue to use endless worksheets and workbook pages to teach math through a dependency on textbook-based materials. These materials are not motivating, nor do they simulate a real learning experience. They certainly do not account for individual differences among students. In fact, for students having learning difficulties, they compound the problem by causing more stress and frustration with the process of learning mathematics skills and concepts.

The teaching profession must "move beyond simplistic, narrow approaches to teaching and learning" (Caine & Caine, 1991, p. vii) and realize that optimal learning does not take place through rote memorization of abstract facts. Furthermore, this narrow focus on facts that are often taught in isolation may even prohibit learning, understanding, and transfer or use of content material by students in today's classrooms. Since many teachers assume that learning can, in fact, take place through memorization of such facts, they seldom do more than provide information and force memorization of these facts and skills. "This is like looking at the moon and believing that we have understood the solar system. There is [much] more" (p. 4). Teachers must capitalize on students' experiences in order for students to be able to form patterns and relationships that are essential for a good foundation in mathematics.

Capitalize on Experiences:
Children live with parallel lines long before they ever encounter school. By the time parallel lines are discussed in geometry, the average student has seen thousands of examples in fences, windows, mechanical toys, pictures, and so on. Instead of referring to the parallel lines students and teachers have already experienced, most teachers will draw parallel lines on the blackboard and supply a definition. Students will dutifully copy this "new" information into a notebook to be studied and remembered for test. Parallel lines suddenly become a new abstract piece of information stored in the brain as a separate fact. No effort has been made to access the rich connections already in the brain that can provide the learner with an instant "Aha!" sense of what the parallel lines they have already encountered mean in real life, what can be done with them, and how they exist other than as a mathematical abstraction (Caine & Caine, 1991, p. 4).

Mathematics, which can be easily related to the real-world, is often taught as a separate subject that is taught one hour, or less, of each school day. In this context, the students cannot possibly be expected to relate mathematics to their own lives. In fact, they perceive the mathematics that they complete in class as work and strive to leave it behind them as soon as the class has ended. Teachers must strive to incorporate "meaningful, challenging, and relevant learning in the classroom" (p. 35) so that all children regardless of learning style preference and background experience will be able to achieve academic success in mathematics.

Another activity that can be used to relate math to the real world involves the use of newspapers. The teacher could cut pages of any newspaper into fourths and give one section to each pair of students. Ask the students to circle any use of numbers, number words, numerals, and the like. Working in pairs or small groups will allow the students to discuss the uses of the numbers as they complete this short assignment. Caution should be exercised to avoid distributing sections of the classified ads since the amount of numbers used would be excessive for the students. Once the students have circled all of the numbers, a list could be compiled to identify the many ways that the numbers were used by asking students to name these uses as the teacher lists them on the chalkboard or on a transparency with the overhead projector. After identifying the uses, the teacher should lead a discussion to identify the math concepts and skills that the students have already mastered and to identify other math concepts and skills that will be covered later in the class. This technique can be modified for primary grades by using sections of the *Weekly Reader* that are used in many elementary classrooms. Through exploring the uses of numbers in the newspaper, students are able to see that math is part of the "real world" and not just some abstract subject that is taught in schools.

Mathematics content and skills instruction can be taught through instructional strategies which allow students to develop as self-confident problem solvers, form appropriate conceptual schemes, and engage in instruction based on diagnosed learning styles and needs (Carson & Bostick, 1988; Jarolimek, 1990). Furthermore, mathematics is the content area most easily adapted to meeting perceptual preferences (modality strengths) of students (Carson & Bostick, 1988). If math is taught in this way, attitudes and achievement may increase.

Of the many competencies that effective teachers must possess, two are adjusting instructional activities to accommodate individual differences (Rupley, 1976) and actively involving the student in the learning task (Fisher, Marliave, & Filby, 1970). Many teachers have difficulty implementing these competencies; however, they can be accomplished through the use of a learning styles model in the classroom.

The purpose of this book, therefore, is to encourage teachers to integrate learning styles based instruction into their math curriculum. By using strategies with which they are already familiar, teachers can accomplish four goals: (a) implement a learning styles model, (b) adjust instruction to individual differences, (c) involve students in the learning task, and (d) ultimately improve student achievement in mathematics.

WHY USE LEARNING STYLES?

Most of us have known for a long time that students and their teachers do not all learn the same way. When given a laboratory assignment in chemistry class, some students start working immediately. They work with the Bunsen Burner until it finally works the way that it should. They learn by doing. Other students start with the instructions. They read the complete assignment and study diagrams before they ever start the experiment. These students learn by reflecting. Other students who need a great deal of structure and learn better under the supervision of an adult will question the teacher concerning all of the details of the assignment and repeatedly ask the teacher to come to their assistance.

Two adults learning to play bridge may approach the task quite differently. One may organize a group to play the game and learn the rules while playing bridge, again learning by doing. The other person may read a book on how to play bridge and the columns in the newspaper on how to bid before s/he ever attempts to play the game.

Many teachers have learned about differences in teaching and learning styles through first-hand experience in the classroom. A student who struggles to understand a new mathematics concept explained in class but cannot comprehend the information may become frustrated by the entire situation. The teacher may sense the interest and commitment of the student and attempt to explain the concept again, but the goal of understanding by the student is still not reached. Sometimes the teacher may feel that s/he could have taught the concept in a better way, wish that the student had received a stronger foundation in mathematics in the earlier grades, or wish that the student applied him/herself more actively in class. However, the only problem that may be present in this situation is that the teaching style used does not fit the learning style of the student. Minds follow different networks to reach understanding, and each different network is simply a different learning style strength or preference.

Extensive data verify the existence of individual differences among youngsters—differences so extreme that identical methods, resources, or grouping procedures can prevent or block learning for the majority of the students (Dunn & Dunn, 1979). Furthermore, strong intuitive appeal surrounds the notion that different individuals learn in different ways. Therefore, different activities or methodologies are required for each student to be successful.

Traditional methods of instruction do not, as any singular method cannot, meet the needs of all students. When students need remediation, time is usually spent reteaching material using the same initially ineffective methods (Carson & Bostick, 1989). The best time to learn the material is when it is first taught; the best way to teach information is to teach it well the first time. In fact, "repetition rarely works; more often than not, it simply reinforces previous failure" (National Research Council, 1989, p. 13).

Worksheets, routine homework problems, as well as other traditional methods of instruction, are usually isolated activities that teach skills in isolation. They are not comprised of instructional patterns that link prior knowledge to the new information being introduced in a meaningful way. "They are generally ineffective as teaching strategies for long-term learning, for higher-order thinking, and for versatile problem solving" (National Research Council, 1989, p. 57).

Some teachers still believe that students who have not learned the material simply have not paid attention in class. There must be a revolution in our actual practices in the classroom relating to how teachers actually teach. Research has continually shown "that most students cannot learn mathematics effectively by only listening and imitating; yet most teachers teach mathematics just this way. Most teachers teach as they were taught, not as they were taught to teach" (National Research Council, 1989, p. 6). In fact, the majority of the failures in schools today is "due to a tradition of teaching that is totally inappropriate to the way most students learn" (p. 2), since many teachers are still teaching in the "pencil-and-paper-era" (p. 63).

In order to implement an effective classroom, teachers must determine the learning styles of students in today's classes. Dunn, Dunn, and Treffinger (1992) describe the effective teacher as one who "varies group lesson plans to accommodate individual students and changes the teaching styles to motivate individual students" (p. 81). The traditional methods of instruction (e.g., textbook, chalkboard, pencil-and-paper activities) do not ensure the active involvement of students, who are, in fact, "turned off" or bored by repeated use of such activities (Carson & Bostick, 1988).

An effective teacher must consider the individual differences of children, adjust instruction to ensure children's success, and provide motivating activities which actively involve children in the learning process. According to Gregorc (1979), gifted and average-achieving students' learning styles may be accommodated by most strategies in traditional classrooms.

However, unsuccessful students may be students whose only problem is that their learning preference is different from the teacher's presentation.

Implementing a learning styles model does not mean restructuring an entire classroom. Through awareness of learning styles, teachers begin to realize that many of the instructional strategies and activities that they currently use in their classrooms are accommodating individual learning preferences. Teachers need only to consider how each of the instructional activities work to accommodate students' learning preferences. This calls for more deliberate, thoughtful planning of activities.

CONCLUSION

Teaching mathematics through individual learning styles can help prevent the perpetuation of the idea that mathematics is "a string of procedures to be memorized, where right answers count more than right thinking" (National Research Council, 1989, p. 10), can promote excellence and equity in mathematics education, can provide various levels of expectations for a diverse group of students, and can establish individualized expectations that are appropriate for each student. Mathematics education for all students must include the expansion of "teaching practices that engage and motivate students as they struggle with their own learning" (p. 58). Incorporating the techniques and strategies presented in this book can enable teachers to achieve the goal of improving student achievement in mathematics.

Effective teachers must consider the individual differences of children, adjust instruction to ensure children's success, provide motivating activities which actively involve children in the learning process, move away from traditional methods of instruction that demand that students work individually to solve specific problems, relate mathematics instruction to the "real world," and remember that unsuccessful students may be students whose only problem in math, or any other subject, is that their learning preference is different from the teacher's presentation (Gregorc, 1979). Furthermore, quality mathematics instruction requires eliminating the obsession with uniformity that is found in many of America's classrooms.

Many teachers are vividly aware of the many complexities of teaching and learning and are knowledgeable of effective teaching strategies including the use of learning styles based instruction. However, many of these teachers have responded to the peer pressure exerted by traditional, sometimes more experienced, teachers. These teachers also have memories of many years of sitting in classrooms where lecture and memorization were the norm, and changing ingrained habits is often difficult. Additionally, many teachers are "evaluated by standards that force them to choose between good teaching and personal survival" (Caine & Caine, 1991, p. 173). To put an end to these counterproductive teaching methods, teachers should be encouraged to "fight a factory model that places a premium on low-quality output and a research model that implies that their observations of what actually happens are invalid" (p. 10). This requires "a strong personality and enormous conviction" (p. 10). Staff development topics for mathematics teachers should include methodology of instruction as well as knowledge of subject matter. Methods of instruction, strategies of classroom management, and findings from current research will broaden the educator's knowledge base. The knowledge gained should empower the teacher to address the variety of diversity among the students. Teachers dedicated to creative teaching which includes a variety of innovative teaching strategies and learning styles based instruction should also search out other teachers with the same aspirations and goals. All of these teachers must continue to change the opinion that narrow or exacting teaching methods can enable all students to reach their academic potential.

Chapter Three

LEARNING STYLES IN MATHEMATICS

HOW CAN I IMPLEMENT LEARNING STYLES?

Before understanding how instructional activities and strategies relate to learning styles implementation, one must have a basic knowledge of learning styles. This may not be a simple process since there are several different learning style models supported by various research and practice. However, there does seem to be a common belief that styles exist and that learning style based instruction can be applied in all areas of the curriculum (Brandt, 1990). Furthermore, "personal awareness is an aspect of all learning style theories" (p. 10). Teachers must realize that they, too, have their own cognitive preferences, the way they best learn new information, and they must also realize that the diverse students in their classrooms have varying cognitive styles. In fact, students' success or failure in a math class may be directly related to the information-presenting strategies and instructional demands of the teacher.

The overall focus on learning styles is on student strengths. Since everyone is different, no one group of students will learn best from any singular methodology or strategy. "Because every child's learning style is unique, every child's giftedness emerges in a different way" (Dunn, Dunn, & Treffinger, 1992, p. 44). Therefore, there is no one best way to teach or learn a given concept or skill (O'Neil, 1990). Perhaps the teachers' most important responsibility is to collect a variety of ways to modify the curriculum in order to effectively teach all students. Many modifications of traditional instruction can be incorporated into classroom strategies to help meet the learning style needs of a diverse student population.

Planning a multitude of different activities for each lesson is not necessary. In fact, two factors contribute to the ease of implementing learning styles. First, more than one area of stimuli (i.e., environmental, emotional, sociological, physical, and psychological) can be accommodated through a single instructional strategy. Second, individuals have a strong preference for only five or six elements (Dunn, Dunn, & Price,

1981). For the other elements, the individual usually has no preference one way or the other. Therefore, a single lesson can accommodate a variety of learning styles.

LEARNING STYLE MODELS

Many teachers do not adhere to any one learning styles model, but use a combination of the ones that most fit the needs of the students in their classrooms. No one learning styles model is necessarily better than another. Teachers must simply find the techniques and strategies that work best in their classrooms. Certain activities and methodologies will work better with some learners while others will not. Through the vast amount of research on learning styles, however, educators do agree that learning styles exist and that accommodating learning styles can produce gains in achievement and have a positive impact on discipline and school climate. Regardless of the label used, educators are realizing and admitting that students have different learning styles that can be used in the teaching process (Wheeler, 1988).

One model of learning styles that facilitates an understanding of learning styles and assists teachers in the implementation of alternative instructional strategies is the model developed by Dunn, Dunn, and Price (1981) and is similar to the one used by Carbo (1992). While the authors of this book do not wish to endorse or promote any particular learning styles model, this model will provide the foundation for providing a brief overview of learning styles by using examples that come from every day occurrences. This overview will give the reader a basic understanding of a learning styles model in order to initiate learning styles based instruction in mathematics.

Environmental Stimuli

The factors relating to the immediate environment of the learner include the need for sound or quiet, the need for cool or warm temperatures, the need for bright or dim lighting, and the need for a formal or informal design. As with all of the factors that will be discussed, the reader should remember that not all elements of learning style are relevant to every student. However, the fact that environmental issues seem to be more complex than previously realized should also be emphasized at this point (Caine & Caine, 1991).

Sound

If one were to look at the environment of most classrooms, bright light, quiet, and a formal room design would be found. This does not fit the learning style needs of all of the students in the classroom. Even though some teachers and administrators feel that students learn more in quiet, orderly classrooms, some students need sound to concentrate. When they are in an extremely quiet setting, they are easily distracted by all of the little noises around them. If a person who needs quiet to study would imagine being in bed an 2:00 A.M., trying to sleep, and hearing all of the noises the house makes which are normally not noticed, then the need for sound may be more easily understood. If students do need sound to study, audio cassette players with headphones can be used while they are working independently in the classroom. The music, classical music or instrumental music, should be supplied by the teacher. Initially this may sound strange to some educators; however, presentations are often given with musical background to enhance reception of the information. Some students even request specific musical pieces while completing individual assignments (Caine & Caine, 1991).

When one school system was implementing learning style based instruction, the students with a need for sound as part of their learning style profile were asked to bring Walkmans® to class. The students only got as far as the door to the school bus before being told that bringing stereos to school was against school policy. The principal of the school and the teachers had to present an inservice session on learning styles to the members of the school board in order to get the rule changed. Once the students were allowed to bring the Walkmans® to class and the teachers provided classical music, the students listened to the music when doing independent assignments in class. This particular school, as well as others, reported increased academic success of students when learning style needs of students were accommodated through instructional practices (Dunn, 1988).

Temperature

The need for cool or warm temperatures is also an element of learning styles which may be part of the learning style profile of students. However, temperature is very relative. One author of this book is very "hot natured" when compared to her best friend, but very "cold natured" when compared to her husband. Even when teachers have a thermostat in their classrooms, making the room comfortable for all students may be difficult. Since some students cannot study if they are too cool and others cannot concentrate if they are too warm, teachers should set the thermostat in the "normal" range and also be aware of "hot" or "cold" spots in their classrooms. Furthermore, if a student is always too cool in one particular classroom, s/he should be encouraged to wear layered clothing so that outer layers can be removed as needed. In other words, common sense should prevail when dealing with this element of learning styles.

Lighting

While bright lights are found in most classrooms, this does not meet the learning style needs of all students. Some students learn better with a dimmer light; they may become hyperactive with bright lights and others may become sluggish under these dimmer lights. Adjusting the lighting in classrooms may not be easy, but turning off one row of lights, if possible, may allow the teacher to meet the needs of more students. Lamps can also be used to create areas with less and more light. These lamps could be purchased at a nominal cost through garage or yard sales and decrease the cost involved from purchasing them from typical commercial sources. The use of cellophane paper or non-flammable crepe paper can also be used to diffuse the bright light. If the lighting fixtures have a tray cover, a rectangle of this paper can be placed inside the tray. Light fixtures that extend from the ceiling can be wrapped with the paper which is secured with staples. Helen Irlen (1991) has noted through research that the bright lights utilized in many classrooms create problems for persons with schtopic sensitivity, a visual perception problem. Diffusing the overhead lights as described or placing colored transparencies over the pages being read can assist in overcoming visual problem.

Design

Typical classrooms also have a very structured seating arrangement with students all seated in student desks or chairs. Children who prefer an informal room design, sitting in a recliner, in the middle of the floor, or lounging on the floor, are very constrained and uncomfortable when placed in traditional student desks. Informal design areas can be created in any classroom by providing a carpeted area with pillows. Sofas, comfortable padded chairs, or bean bags also provide for an informal room design. Additionally, allowing students to contribute their own pillows and other appropriate items helps to develop a sense of ownership in the classroom, as well as aid in designing the appropriate floor plan. Students, as a group, can aid in the decision process of arranging furniture, structuring breaks, and decorating bulletin boards and walls (Marzano, 1992).

If a teacher chooses to implement informal design as part of learning styles based instruction, arranges a section of the classroom with bean bags, study pillows, or other less formal furniture, and tells the students that they may choose this area for work, the majority of the students may want to sit in the new area the first day. Classroom management guidelines should be clearly established which include staying on task, completing assignments, not disturbing neighbors, and the like. Once a student violates any of these rules, s/he should be promptly returned to his/her desk for a few days or until the student outlines a plan for his/her return, which places responsibility for abiding by the rules on the student. Furthermore, after approximately two weeks, the students who do not need an informal design will return to their desks through their own initiative. Only approximately 20 percent of the classroom population has the need for an informal design area in their learning style profiles (Dunn, 1988).

Prior to learning about the importance of learning styles, one author of this book made her son sit at the kitchen table to do his homework. The result was many arguments, not better grades. Now her son is allowed to study wherever he chooses. In an informal environment such as reclining on the stairs or in the middle of the hallway, he completes his homework with fewer arguments. Administrators and teachers often report similar results in their classrooms and may even wonder why this has not been used previously since they usually sit on the floor at home when completing serious work.

Emotional Stimuli

The second group of stimuli includes emotional elements. Emotionality affects a person's learning style through motivation, persistence, responsibility, and structure. Through a learning styles model of instruction, these factors can be accommodated for all students, regardless of learning style preference.

Motivation

Motivation is often defined as a person's desire to achieve and usually relates to a student's desire to receive a high grade (Caine & Caine, 1991). It can change rapidly from teacher to teacher, class to class, and subject to subject. Motivated students are interested and eager to learn. Students have increased comprehension when they are motivated to learn mathematics concepts and skills (Fass & Schumacher, 1978); and since all students are not motivated to learn mathematics, the teacher's responsibilities include trying to increase motivation in the mathematics classroom, and, therefore, igniting the personal curiosity of each child. Kutz (1991) suggests other alternatives for motivation enhancement such as having a "productive, and lively mathematics classroom" (p. 11) and "making connections and showing relationships" (p. 19). Additionally, "invoking the power of play" (Caine & Caine, 1991, p. 138), modeling an enthusiasm for mathematics, and incorporating the provision of choices, novelty, and curiosity also address the elements of motivation and persistence. If mathematics instruction is based on real-life experiences and the students realize that specific mathematics concepts and skills are related to goals in their own live (meaningful to them in a personal manner) they will be more motivated to learn the material. However, the teacher must realize that the goals of the students may range from a career in aerospace to receiving the correct amount of change from a $5.00 food-stamp coupon at the grocery store. Some students are simply harder to motivate than others and any goals that are incorporated must be very individualized and specific to each child's "real world" expectations. One way to assist the students in understanding the "real world" uses of mathematics is to have them report how they and their families have used mathematics away from school. This activity can be accomplished through personal reflections of the students in journal writings or by having the students question parents and other adults and report the findings orally in class.

Persistence

Persistent students stay with a task until it is completed. Others need breaks; they cannot continue if they are prohibited from taking breaks. When these students who lack persistence experience any difficulty, they lose interest, become irritated, and begin to daydream. These students should be allowed to complete short tasks or assignments with breaks between them; they do not like to work on one long, extended project or assignment. Mathematics teachers having students who need breaks can try the following persistence strategy.

Persistence Strategy:

If an assignment would normally include 25 problems, the teacher should first model the procedure for solving these problems using manipulatives (see Chapter 4), calculators, or other appropriate instructional resources. The teacher's demonstration would include various problems which model the concept or skills being taught, as well as the first problem in the class assignment. Then the assignment is divided into quarters making it four small assignments of six problems each (Dotson, 1988). The following choices are given: (1) choose any four of the six problems to complete, (2) work the problems independently and check your work with a calculator, when appropriate, or (3) work with a partner at any of the following learning stations: the overhead projector, chalkboard, or a small table. Once the students have completed the first assignment, they take the work to the teacher to be checked for errors or simply give the assignment to the teacher if the problems are more complicated and require closer analysis by the teacher. When the papers are given to the teacher, smaller pieces of paper may be needed to reduce the amount of paper used. Regular notebook paper can be cut in halves or fourths, depending on the amount of space required for the assignment. The difference here is that the students go to the teacher instead of the teacher going to the students. Once the first assignment is completed, the second assignment which includes choosing any four of the next six problems, is completed in the same way.

Short breaks between assignments may encourage less persistent students to perform better in mathematics. Furthermore, completion of the first assignment will create a successful experience for the students, success will set the foundation for more success, and these students may also be more motivated to finish the remaining assignments. Having a choice regarding which four problems to work will also be motivational to some students. They feel that they have permission to omit the more difficult problems when, in fact, all of the problems are on the same level of difficulty.

Responsibility

Responsible students follow through on a given task, complete it to the best of their ability, and often do so without direct or frequent supervision. If responsibility is defined as doing what "ought to be done," it could also be defined as conformity. Students with low responsibility are non-conformists (Dunn, 1988); they want to do things their own way and don't want to do what they are told to do. These students often permit their attention to become diverted when a task becomes difficult.

Responsibility Strategy:
With nonconforming students, teachers should do three things: (1) let the students know the teacher believes the task is important; (2) talk to the students like a colleague; do not talk down to them, dictate, or use authoritarian teaching styles; be diplomatic; and (3) give the students a choice of how they will demonstrate the learning (Dunn, 1988).

Basically, these students need collegial teachers, not authoritative teachers who mandate what should be done. These students have repeatedly been told that they need to complete assignments because this will make them more educated and will prepare them for higher level mathematics classes. They have not been motivated by this approach and may be more willing to do the assignment when they realize the task is important to the teacher. There is a certain amount of motivation to please the teacher in students of all grade levels, IF the students believe that the teacher truly cares about them. Instead of telling the nonconforming students to start the assignment NOW, the teacher should tell them the purpose of the assignment and ask them to take a few minutes to get organized and to decide which option they will choose. The options given to the students might include working the mathematical problems in the traditional pencil-and-paper way, demonstrating how to work the problems to a small group, designing a poster or bulletin board which explains the process, producing an electroboard which involves the assigned concept, building a model which represents the concept, or producing an audiotape or videotape which explains the concepts and uses them in a creative way. These nonconforming students will be more likely to attempt the assigned task when approached using this strategy.

Structure

Structure is simply the establishment of specific rules for working on and completing an assignment. It determines the number of options available to the student. When students tend to ask many detailed questions about an assignment, they very likely have a high need for structure. They find it difficult to achieve without the imposed structure. Other, more creative students squirm under mandated guidelines and find learning that way frustrating and unstimulating. These students have a low need for structure. Classroom management for learning styles based teaching requires a great deal of structure; however, there is freedom within the structure. Choices are allowed, but the decisions made by the students are limited to a few options.

Increasing Emotionality Elements

The primary factor in accommodating children whose learning style preferences are low motivation, low persistence, low responsibility, and a great need for structure is providing material that is interesting and meaningful and showing children how the material can be used in other situations, as discussed earlier. Students with these learning preferences need to be prepared for understanding the presented material, and the delivery of the curriculum should include a variety of perceptual methods. That is, the teacher must provide background knowledge, give a purpose for learning the information, concept, or skill, provide for manipulative experiences, and divide extensive concepts into shorter units or subskills. Simply by giving students this guidance, teachers decrease the effort that the students must exert to understand the material and increase these emotionality factors (Irwin, 1990).

Mapping (or webbing) strategies can be used to provide structure and organization for students who need it. Mapping can be used to help students read, understand, and compute answers to word problems in math. The word problems used in this type of activity should be meaningful, interesting, relate to "real world" situations, and be developmentally appropriate.

Mapping Strategy:

The teacher could provide a visual outline for the story problem in mathematics and discuss the outline to give the students key points to find in their reading; the students read the word problem; the teacher and students fill out the map together; and the students compute the answer to the word problem. This visual outline could be in the form of a graph or table, a problem solving strategy taught in many classrooms.

Mapping also provides students with motivation by giving them a task for which they can see an end and by providing structure that can be used in future learning. The method described assists students in staying on task, being involved, and completing the task with little assistance.

Sociological Stimuli

The sociological elements concern grouping students for instruction. There is no best way to group students for maximum learning. Students learn in a variety of sociological patterns that include working alone, with one or two friends, with a small group, with adults, or in any combination of these patterns.

Learning Alone

For students whose learning preference is to learn alone, many of the traditional instructional methods, such as workbooks and worksheets, are appropriate as far as the sociological preference is concerned. However, these methods are not effective for ensuring application of these skills in other situations. There are other instructional activities that will accommodate the learning preference and ensure active involvement in the learning task. Mapping (or webbing) strategies can be effective strategies for students who prefer learning alone.

Group Learning

Strategies for students who prefer to learn in groups and whose preference is learning in combined ways can be accomplished using cooperative learning. When cooperative learning activities are used, each member of the group should have an assigned task to ensure active participation by the whole group. These tasks could include recorder (writing the results of the assignment), reporter (reporting the group results to the whole class), time keeper (making sure the task is completed within the specified time limits), and cheerleader (rewarding members for their participation and ensuring participation of all members of the group). The assignment of roles provides experiences in accepting responsibility within a group setting.

While heterogeneous groups are advantageous, caution should be used when grouping academically achieving students with average and lower achieving students since these academically talented students may prefer to work alone or in a group of their true peers (Dunn, 1988). If these students are placed in truly heterogeneous groups and a grade is assigned for the cooperative effort, these students may become so concerned with the grade that true learning is diminished. Instead of grades being assigned for the group work, rewards can be earned. The students will later take a test to receive individual grades. Underachieving or slowly achieving students may prefer to work in small groups and may be anti-adult in their sociological preference (Dunn, 1988). Therefore, the use of cooperative groups will be advantageous to these students, but the use of one-on-one remediation with the teacher should be limited, if used at all. The implementation of cooperative learning strategies into a classroom should address the sociological preferences of all students.

Variety of Sociological Groupings

A class of students who learn in a variety of sociological patterns that include working alone, with one or two friends, or with a small group may be easily accommodated by giving students the option of working in any of these sociological grouping patterns. The following example is how one secondary math teacher decided to give choices on assignments.

Choices In Grouping:

For in-class and homework assignments, students can choose to work alone, in a pair, or in a small group of no more than four members. Regardless of the sociological grouping chosen, all students turn in the same, completed assignment. However, each person must check and sign-off on all other group members' work before giving it to the teacher to be graded. While all members of the group receive the same grade for these assignments, each student is tested individually.

After the teacher implemented this system of grouping choice, two boys in the class who had been failing math were receiving B's by the end of the semester. One of the boys went on to successfully complete an advanced math class the next year (Dotson, 1988).

Physical Stimuli

Physical stimuli include perceptual preferences (i.e., senses through which individuals receive information), time of day, intake, and mobility (Dunn & Dunn, 1979). As with other learning style elements, all of these will not be relevant to all learners, but the modality strengths or perceptual preferences are important for a majority of students in today's classrooms. Specific activities which accommodate a variety of learning style elements including perceptual preferences are presented in Chapter 6.

Perceptual Preferences

Seeing, hearing, and touching or feeling are most often thought of as the senses of learning; whereas, visual, auditory, tactual and kinesthetic are thought of as the modalities of learning or perceptual preferences. Students will tell teachers through their daily actions what their modality strengths are. Based on this knowledge, teachers should provide activities to accommodate all four modalities as each mathematical concept or skill is taught.

Visual Learners. Students who are visual learners need to see the information and prefer to watch demonstrations. These children would rather read than have the material read to them. They are meticulous, like order, and are easily distracted by movement. These are the students who quickly notice any visual changes in the classroom, a new bulletin board, poster, or display or the teacher's new glasses, hair style, or new clothes. These students remember what they have seen, not what they have heard. Resources to use with visual learners include bulletin boards, task cards, charts, posters, videotapes, flash cards, overhead transparencies, and slides, as well as the traditional chalk board. (See Chapter 6.) Visual learners will say "show me what you mean" when they do not understand the concept being discussed (Carson & Bostick, 1988).

Auditory Learners. Students who are auditory learners need to hear the information and learn through auditory repetition. They are constantly talking and want to discuss new concepts. Children who learn through the auditory mode talk themselves through difficult problems and often move their lips while reading silently. Because of this often incessant talking, they are easily identified by the teacher. They learn about their world by hearing and talking about it. Questions such as, "How do you know that your answer is correct?", "What did you do to get that answer?", or "Do you agree with the answer given by _____ — Why or why not?" are useful in directing the responses of the auditory learner. Having these students discuss the mathematical process that they used is also an excellent strategy to use. Other resources appropriate for the auditory learners include the traditional lecture, audiotapes, cooperative groups which discuss the concept, and rhymes, raps, or jingles which include content specific information, as well as the question and answer activities (see Chapter 6). Auditory learners use words such as tell, say, discuss, explain, and listen and will say "tell me what you mean" when shown a visual display (Carson & Bostick, 1988).

Kinesthetic And Tactual Learners. Both kinesthetic and tactual learners need to be physically involved in the learning process. Learning style theorists may refer to the perceptual strengths as being tactual/kinesthetic (T/K) or the characteristics may be separated into the two categories, tactual and kinesthetic. The tactual learner needs to touch to learn, while the kinesthetic learner needs more involved body movement as an aid to acquiring the information. Both of these types of students learn through hands-on activities and direct involvement. They remember best what they have done, not seen nor heard. Tactual or kinesthetic learners may be poor spellers and write words to see if they "feel" right. Some good resources to use with these learners include experiments, task cards, eletroboards, puzzles, large floor games, manipulatives, and models (see Chapter 6).

Tactile Learners. Tactual learners utilize fine muscle skills in acquiring knowledge. Doodling and paperfolding are favorite pastimes observed in a classroom setting. Constructing models which are built of clay, toothpicks, sugar cubes, styrofoam pieces, and the like are representative of work enjoyed by the tactual learner. Manipulatives (see Chapter 4) such as counters, pattern blocks, beads, calculators, Cuissenaire rods, and base ten blocks are excellent tools to allow these students to "feel and experience" mathematical concepts.

Kinesthetic Learners. Mobility is a high need for the kinesthetic learner. Strategies involving more body movement while academic tasks are performed assist the kinesthetic learner to a great degree. Examples include jumping rope, dribbling a basketball, rocking in a chair, walking or riding a stationary bicycle. Instead of saying "tell me what you mean," they will usually ask to try to complete the task involved; they want to do it themselves (Carson & Bostick, 1988). These students are also quite easily identified in the classroom due to their high need for movement.

These students usually have the shortest pencils in the classroom for two reasons; (1) they learn quite early in their educational career that going to the pencil sharpener is an acceptable form of movement in many classrooms; and (2) they bear down harder on the pencil than other students in order to try to get more physically involved in the assignment. This increased pressure on the pencil results in the lead breaking or the lead being used rapidly and the pencil requires sharpening for further use.

Mixed Modality Strengths. Some students may have mixed modality strengths. They have two or three modalities which are equally efficient in perceiving and retaining the information. Regardless of the modality through which the material is presented, they can process the information in an efficient way.

Observing Modality Strength Characteristics. As stated earlier, students will tell teachers through their daily actions what their modality strengths are. Educators only need to observe them more closely. A third-grade teacher, who used learning styles in her classroom, returned corrected spelling papers to her students. The students kept them for approximately two minutes. After returning the papers to the teacher, the students were told to write the words that they had misspelled. The visual learners wrote the words exactly as they had misspelled them on the test. They had looked at their grades and the words they misspelled, and the incorrect spelling was reinforced. Especially with visual students, teachers should take the time to write the correct answer. These students will be more likely to remember correct answers, not incorrect answers, if this simple step is taken.

When auditory students were told to write the misspelled words, they immediately started to complain, saying things such as, "We didn't look at the ones we missed," "I know my score, but I don't know which words I missed." They had not looked at the words they missed because of their preference for hearing information. These students complained that they had not been told to look at the words they had spelled incorrectly.

There were two tactual/kinesthetic learners in this class. When asked to return the papers, the first student had already wadded the paper and thrown it in the trash when the teacher asked for the paper. The other student had folded his paper a number of times, put it in a book, and put the book in the desk. He looked through all of his books and his neighbor's books, but he never found the paper. Folding the paper involved small muscle movement (tactual) and wadding the paper and moving to the trash can involved full body movement (kinesthetic).

Teaching Through Perceptual Preferences. Regardless of the perceptual preferences or modality strengths of the students, ALL students should have the opportunity "to be engaged in talking, listening, reading, viewing, acting, and valuing" (Caine & Caine, 1991, p. 6). In short, all of the senses should be used when activities and resources are designed by teachers in order to provide every student with a variety of sensory experiences that deal with the mathematical concept or skill being taught. Once the information, content, or skills have been introduced in the math classroom through, hopefully, the students' perceptual preferences or modality strengths, the information should be reinforced through an additional modality of learning, and, finally and most importantly, the students should be able to use the information in a new and creative way. This gives the students the opportunity to detect patterns and connect these patterns to prior learning in the math classroom. For example, allowing students to play "around with a formula" (p. 8) may allow the students to develop a better appreciation for the formula and what it really means than students who can memorize the abstract fact (the formula) but not be able to "manipulate it creatively" (p. 8). Additional creative uses of new information include allowing students to make electroboards, learning circles, or audio tapes, to model the desired learning for the class, to interview authorities, and to create other learning games (see Chapter 6).

Regardless of the modality strengths of the students, activities that are fun, rewarding, and nonthreatening should be used to make mathematics more meaningful. For example, the concept being studied could become the theme for a designated period of time.

Understanding Division:
When beginning to teach the concept of division in the math classroom, make division the theme for the entire day or week. When this happens, every activity throughout the day will relate to division in some way. Parents might bring pizza to school for lunch and each pizza must be **divided** into equal parts for all of the students. Time on the computer or working at a favorite learning center must be **divided.** The areas in the room may also need to be **divided** for various activities and projects throughout the day. The teacher constantly talks about division,

the students see division, and the students may even be **divided** into groups to "feel" division.

Through these strategies and techniques, the students truly experience division through all modalities of learning. This type of organization of activities also helps to relate the concept of division to the real world and makes learning more meaningful for the children (Caine & Caine, 1991).

Intake

Intake, another physical element of learning styles, is a need for individuals who often nibble, drink, or chew gum while studying. Three possible reasons for the need for intake include food being sought to replace the energy being expended, intake may relax the tension that some people experience when concentrating, or the constant need for intake may simply be the result of poor eating habits. However, some students who are diabetic or hypoglycemic may have medical reasons for needing several small snacks a day instead of the three traditional meals. Students who continually bite their fingernails or pencils may be signaling a need for intake. Since this need is only apparent in six percent of the student population (Dunn, 1988), it is not a major concern for most teachers. Labels on purchased snacks should be critiqued for sugar content. If sugar is listed as one of the first five ingredients, then that product may not be appropriate for classroom consumption. However, allowing students to have a cup of water or celery sticks during class may accommodate the need for intake in their learning style profile.

Time of Day

Another physical element which affects students and adults alike is the time of day preference. Most adults can easily identify the time of day when they are most productive. Some people do their best work early in the day while others are more productive late at night. As students progress through school, they tend to adjust to the time of day factor. However, they may still have a time of day when they are more productive. The peak period for most elementary school students is from 10:30 A.M. until 2:30 P.M. (Andrews, 1991; Dunn, 1988). Reading, a subject which is considered critical in the elementary school, is not usually taught during this time frame. When one elementary school in North Carolina tested the fifth graders for learning style including time of day preference, 62 out of the 67 students tested were identified as afternoon learners. As a result of this testing and the mismatch between time of day preference for the majority of students and the time that reading and math were taught, the daily schedule was reversed so that reading and math were taught later during their peak period. The administration of the school system reported a miraculous change in attitude and marked increases in reading and math achievement (Andrews, 1991).

Mobility

Some students, particularly the kinesthetic students, need the freedom to get up and move about when studying. They cannot function well unless they are permitted to vary their posture and location. Those students who need mobility will perform better academically if they have the opportunity to move about at certain intervals. Short assignments at various locations in the room will aid students who need mobility, as will the strategy suggested when discussing persistence.

Psychological Stimuli

The final grouping of learning style elements is psychological stimuli. Psychological stimuli include global verses analytic learning, right and left hemisphericity, and impulsive verses reflective thinkers.

Global Learners

Global learners process information in a whole-to-part format. They need to see the entire picture before they concentrate on the smaller steps. In other words, they need to know where they are going before they find out how to get there. These people need to grasp the entire concept before they are ready to see the details (Dunn, 1988). "Global learners need frequent breaks, concentrate against a background of music or conversation, and enjoy soft lighting and informal settings (Dunn, Dunn, & Treffinger, 1992, p. 20). In mathematics, for example, they need to see the long division problem completely worked before they are ready to take the process apart and put it back together in a step-by-step process. Other techniques to use when teaching global learners include starting the lesson with a short story, myth, or an antidote to get their attention and enable them to organize the following information in story form to aid memory. Drawing shapes around key facts, and using different colors of chalk or transparency markers for different steps in mathematical problems are also effective strategies to use with these students.

They often learn through jokes, fun, graphics, symbols, riddles, puzzles, humor, drawings, and visualizations. Simply starting with the facts often results in lack of attention by these students.

Analytic Learners

Analytic learners, on the other hand, learn best when information is presented in a step-by-step process; they learn sequentially through one fact at a time to build understanding of the concept being presented. Since mathematics is usually taught in this way, it is usually easier for the analytic learners to understand than global learners. There appears to be a mismatch between the way teachers teach and the way students learn since 65 percent of all teachers are analytic and teach sequentially and most students in kindergarten through middle school are global learners (Dunn, 1988). The present curriculum can be adapted for the various type of learners through the inclusion of different methods and techniques of instruction and the addition of manipulatives.

Left Hemisphericity

Language, specialized verbal expression, and mathematics skills and part-to-whole processing tend to be left-brain functions (i.e., analytic learning style). Left-brained learners tend to be more verbal, abstract, and logical; they also tend to be very time and product-oriented, and are basically very rational people (Caine & Caine, 1991). They are often auditory learners and make use of deductive reasoning (Dunn, 1988). Furthermore, Cherry, Goodwin, and Staples (1989) report that the traits of the left hemisphere are more easily measured than the right hemisphere traits since language skills are associated with the left hemisphere.

Right Hemisphericity

Visual art and music skills, along with physical coordination and whole-to-part processing tend to be right-brain functions (i.e., global learning style). Right-brain learners are visual learners who tend to be intuitive and holistic. They are divergent thinkers who remember things through pictures and often find it difficult to express themselves verbally. However, they have excellent spatial recall (Caine & Caine, 1991; Dunn, 1988). They tend to learn better through tactual or kinesthetic resources, may not remember what they are told, and, therefore, may have trouble in conventional classrooms (Wheeler, 1988).

Teaching to Both Hemispheres

Left-brained students may learn more about dinosaurs from a lecture than a visit to the museum, while right-brained students may learn more about dinosaurs by visiting the museum and experiencing the learning first hand. When a field trip to the museum is impossible due to financial and time restraints, a videotape can provide the initial overview required by the global, right-brained students. The discussion which should follow the viewing would accommodate the analytic, left-brained students, and therefore, both groups of students would be accommodated by the same lesson.

School curricula tend to focus on left-brain functions or the analytic learning style. Linearity, analysis, and thinking sequentially are emphasized. Typically, imagination, wonder, and creativity, which are right-brain functions, are inhibited. However, no person uses all one side of the brain. That is, left and right sides of the brain work together; they are both involved in every activity that we encounter daily. In other words, many "left-brain" tasks are enhanced through "right-brain" processing and vice versa (Caine & Caine, 1991). Think back to first grade or kindergarten; you probably learned the alphabet through the ABC song. The ABC's are sequential, analytic information and the song is more right-brained. This is only one example of the two sides of the brain working in harmony. Furthermore, students "who use both sides of the brain equally [tend to] fare best academically and socially" (Webb, 1983, p. 514). Cherry, Goodwin, & Staples (1989) offer the following implication for educators:

> New experiences may evoke one bit or the whole pattern of one type of information. Thus, it is important that children constantly have new multisensory experiences, presented in a variety of ways, in order to build their recall and memory. Because different techniques will trigger the storage and the recall of memory differently in different individuals, it is imperative that many approaches be used (p. 27).

Impulsive And Reflective Thinking

Impulsive verses reflective thinking is the remaining psychological element. Impulsive learners make snap decisions, but reflective thinkers will consider all of the possibilities before they make a decision (Dunn, 1988). However, a certain amount of reflection is necessary for any higher-order thinking and just as no one is totally right brained or totally left-brained, no one is totally impulsive nor totally reflective.

Since reflection is a powerful tool, activities which encourage students to reflect upon their own experiences that relate to mathematics should be encouraged. One example of this is journal writing, which is a "powerful way to process experience" (Caine & Caine, 1991, p. 153). Teachers should have a reflecting time on a regular basis and encourage the students to write about personal experiences or ways that they have used math in their individual lives. This not only allows reflection to occur, but also helps the students to see the necessity and value of mathematics. In addition to journal writing activities, other activities which increase reflection include composing personal definitions of mathematical terms instead of reciting the textbook definitions, writing explanations of how to solve particular problems, or creating word problems based on specific criteria.

Similarities in Psychological Stimuli

There are many similarities between impulsive thinkers and global learners and reflective thinkers and analytic learners. The characteristics of each of these learning preferences tend to overlap.

Simultaneous Learners. Students who learn best through global techniques, who are right-brain processors, and who are impulsive thinkers tend to learn best with an overall picture of the concept (Dunn, Dunn & Price, 1981). These students, as stated earlier, are simultaneous learners. Teachers should redirect these students' attention by asking how things make sense within the entire passage in the word problem or concept in mathematics and emphasize relationships of specific facts to overall meaning (Walker, 1988). Additional strategies that can accommodate these learners include guided imagery, semantic maps, story maps, and color coded division problems to identify the specific steps involved in solving the problem. Having the new type of problem worked first and then demonstrating the techniques of solving the problem allows the student to view the end product before being given the "parts" of the solution process. All of these strategies guide the reader to examine the whole text or content, rather than first examining separate ideas and then putting them together.

Successive Learners. Students who process information in a step-by-step process are referred to as successive learners (Walker, 1988). These students learn best through analytic techniques, are usually left-brain processors, and are reflective thinkers (Dunn, Dunn, & Price, 1981). Any method in which the teacher models what is expected would work well for these students. Teachers should utilize the "think-aloud" techniques developed by Madeline Hunter (1989) and Roger Farr (1990). The teachers verbalize the processes and thoughts that go through their minds when determining how to work a specific type of math problem. In other words, the teacher models or guides the thinking process sequentially.

UNDERACHIEVING STUDENTS AND LEARNING STYLES

According to O'Neil (1990), at-risk students, those with an increased chance of failure in school due to "personal behaviors, past educational records, or family problems" (p. 5), have more to gain from alternative methods of instruction based on learning style needs. The overuse of traditional methods of instruction and the lack of alternative methods often work against these underachieving students.

Dropouts, students who "can't make it" in school, have learning style needs that are not accommodated by the current system of education (O'Neil, 1990). These students tend to work better in soft light and an informal room design such as sitting on the floor; they prefer to work in pairs or groups, have a high need for movement, are not motivated, responsible, nor persistent, learn best through tactual or kinesthetic activities, and tend to be global and/or impulsive learners. These students are often labeled as underachievers or problem students since their learning style profile does not match the teaching style of most classrooms. While many teachers are promoting quiet, independent studies, as discussed earlier in this chapter, these students learn better through direct experience, cooperation, collaboration and interaction (O'Neil, 1990).

Sometimes teachers learn the importance of incorporating tactual/ kinesthetic strategies when teaching "slow learners" quite by accident. One second grade teacher of remedial students was trying to teach an English lesson on commas, periods, and exclamation marks and what each represent. After several explanations, an auditory activity, she had the students read aloud. She expected the students to slow down when they saw a comma, pause when they saw a period, and read with enthusiasm when they saw an exclamation point. However, since the teaching methodology did not match the learning style profile of these underachieving students, they paid no attention to the punctuation marks as they read the passage. Finally, in desperation, the teacher used the following strategy (Caine & Caine, 1991).

Underachieving Students Learn Punctuation:
The teacher had the students walk around in a circle, when she said "comma," they slowed down; when she said "period," they stopped; and when she said "exclamation mark," they jumped up and down. After trying this simple technique for five minutes and then having the students read the passage again, all of the students slowed down when they saw a comma, stopped when they saw a period, and read with enthusiasm when they saw exclamation marks.

This teacher finally taught these children in a way that made sense to them, through their individual learning styles. These underachievers were tactual/kinesthetic learners with a high need for mobility.

In addition to being supported through everyday classroom experiences discussed by teachers, the fact that learning style based instruction is especially appropriate for academically at-risk students is also supported by a variety of research. This research reports that positive effects abound when learning-style based instruction is used with special education students, under-achieving populations, students in low socioeconomic areas, minority students, and students who have experienced "traumatic family upheavals" (Dunn, 1990, p. 17).

CONCLUSION

Effective teachers must consider the individual differences of children to capitalize on student strengths in mathematics instruction. Students should, in fact, be allowed to work through their individual learning styles whether this includes working in groups, discussing mathematical concepts in a variety of situations, or presenting and sharing key ideas with their classmates, in order to "take charge of their own learning" (National Research Council, 1989, p. 59). Since student learning styles are as unique as fingerprints, instruction must be individualized as much as possible to increase the students' understanding of various mathematical concepts, as well as how they can best learn new content through their own learning styles.

However, implementing learning styles may seem like an impossible task. The reader may now be wondering, "How can I do all of this?" First, any teacher wishing to implement a learning styles approach in mathematics, or any other curricular area, should start with only one or two elements and gradually add other elements. Teachers must remember that each student is unique and that using learning styles can help students become life-long learners. Through implementing learning styles based instruction, teachers can create learning environments that are suited to the needs of a diverse group of students, include team-work and group problem-solving strategies in mathematics lessons, use approach and speed to adjust instruction to individual needs and abilities, and, in turn, increase all students' chances for success in mathematics.

Teachers are encouraged to talk to colleagues that have experienced success through implementing a learning styles approach to instruction, because teachers remain the key to establishing effective instructional techniques in math classrooms. The transition will not be an easy one, but it is well worth the expended effort. In fact, the journey can be "exciting, challenging, and immensely rewarding" (Caine & Caine, 1991, p. 180). Furthermore, as we travel along this fascinating journey, we will be proud to say, "I gave them my best because I AM A TEACHER!"

Chapter Four

USE OF MANIPULATIVES
FOR INCREASED COMPREHENSION

NEED FOR CONCRETE EXPERIENCE

All students, regardless of learning style preference, age, or grade level, must have a good foundation in mathematics. Without a solid foundation in mathematics, students will continue to experience confusion, frustration, and stress when dealing with mathematics in their lives, and will continue to wander in the symbolic world of adults without ever finding their way through this complicated maze. Students must be given personal, concrete experiences which fit their learning styles to enable them to find their way out of this seemingly endless stream of difficult concepts. The lack of a good foundation in mathematics is evident in the preschool child who exhibits her ability to count to ten by saying, "one . . . two . . . three . . . seven . . . nine . . . ten." This student is enthusiastic and proud of the fact that she is learning to count. However, she has not yet developed a foundation nor an understanding of what counting actually means. There is no reason why nine comes before ten. In order for the child to obtain the necessary foundation, real objects must be counted, each of the numbers must be explored through a variety of concrete objects, and a variety of groupings for each number should be discovered. This is an essential first step in learning to count. The use of manipulatives to provide concrete experiences for students is necessary regardless of the concept being presented. In fact, mathematical skills and concepts should always be introduced through concrete experiences, whether students are visual, auditory, tactual, or kinesthetic learners, kindergarten students, sixth graders, or high school students.

The kindergarten student in the above counting example was excited about learning mathematics. When given the opportunity to explore patterns and relationships through a wide variety of concrete objects, most young children will be motivated by a natural curiosity. However, as they progress through our nation's educational system and mathematics is taught through abstractions and repeated pencil-and-paper activities, these same students "begin to view mathematics as a rigid system of externally dictated rules governed by standards of accuracy, speed, and memory" (National Research Council, 1989, p. 44). A subject that was once motivating and interesting to them becomes frustrating and boring. Furthermore, this method of teaching may fit the learning style profile of analytic learners who learn best when the information is presented in precise steps, but this type of instruction does not meet the needs of right-brain, global learners. Mathematics teachers should concentrate on developing teaching strategies that capitalize on children's natural curiosity, engaging students in learning, motivating students as they learn mathematics in a more natural way, and remembering to consider the students' learning styles as each lesson is being planned.

While some educators criticize the use of concrete materials in the classroom because it is seen as idle playtime, this exploration of materials enables the students to work through various situations that foster logical thought. Both Piaget and Brunner have emphasized the need to use concrete items and manipulative experiences when teaching mathematics to children. Furthermore, the use of concrete manipulatives in the mathematics classroom closely adheres to the constructivist theory, which is one of the two widely accepted mathematics learning theories. The constructivist theory emphasizes student involvement and participation in learning. The students interact with and manipulate concrete materials so that mathematical knowledge and concepts develop naturally over a period of time. Learning is individualized and based on each student's learning style and previous experiences. However, very few mathematics teachers base instruction on the discovery theory of learning even though the ideal is attainable. The main reason for the lack of instruction based on the constructivist theory is the fact that the other theory of mathematics learning is more popular and much easier to implement (Kutz, 1991).

The other widely accepted theory of mathematics learning is the transmission theory. Teachers who base instruction on this theory provide or present information to the students and encourage rote memorization of facts (Kutz, 1991). The students are often passive learners who remember only enough facts to pass the test and forget the information as soon as the test is over. Obviously, a classroom based on the transmission theory is very limited in meeting the diverse learning styles found in most classrooms. At best, it may enable auditory learners to retain information regarding basic facts and steps completed in a problem, but seldom is true understanding of the concepts achieved by any of the students regardless of learning style. Unfortunately, the transmission theory is the one used by many of today's mathematics teachers. Although presenting information to the students and encouraging rote memorization is considered by a number of authorities to be the least effective methodology for teaching mathematics (Baratta-Lorton, 1976; Caine & Caine, 1991; Carson & Bostick, 1988; Kutz, 1991; National Research Council, 1989; and Willoughby, 1990), instruction based on the transmission theory "prevails in most of America's classrooms" (National Research Council, 1989, p. 57). "It has [even] been said of American education that people learn by doing but teachers teach by talking" (Willoughby, 1990, p. 94). This type of instruction does not promote long-term learning, higher-order thinking, nor problem solving skills.

The understanding of patterns in mathematics through the use of manipulatives will enable students to "see relationships and interconnections in mathematics and . . . enable them to deal flexibly with mathematical ideas and concepts" (Baratta-Lorton, 1976, p. xiv). Any worksheet or workbook page simply cannot involve the student in real-life experiences, nor accommodate a variety of learning styles as well as concrete materials can. Above all, basing instruction on the transmission theory does not allow all students to succeed in mathematics since this method does not match the learning style needs of most students. The use of manipulatives can involve all of the modalities of learning; these materials can be seen and felt by the children as they talk about them with other classmates and the teacher. If abstract symbols are introduced before the concept has been developed through the use of concrete experiences, these symbols can impede understanding instead of enhance learning. In other words, the abstract symbols should be used only after the students have developed a basic understanding of the concept or skill. Teachers, although they often do, should never introduce any new math concept or skill simply by writing the math problem (abstract symbols) on the board and demonstrating how the problem should be worked.

All math concepts and skills "can, and should, be derived from something that is real to the learner" (Willoughby, 1990, p. 11). This reality can often be concrete items from the students' past realm of experience. However, many teachers believe that concrete materials should only be used with slow learners or underachievers and that average or above average students in math do not need these resources. When these materials are not used with these students, they may be able to grasp the abstract concept or skill, but seldom will they be able to relate mathematics to their everyday world. Eventually these mathematically-achieving students could lose interest in these manipulations of abstractions and will choose not to take higher level math courses. "By deriving mathematics from the learner's reality and by constantly applying that mathematics back to real situations in which the learner is interested, we can help students to understand mathematics better and to see it as a useful, powerful, and even beautiful tool that helps them solve their problems and helps them understand the world around them better" (Willoughby, 1990, p. 11). Furthermore, if teachers use manipulatives and concrete experiences to introduce and develop mathematical concepts and skills, more of the "underachievers" in mathematics will come closer to reaching their academic potential in this content area.

While classrooms that incorporate use of manipulatives, concrete objects, real-life experiences, and active learning take a great deal more time and effort on the part of teachers, the result will be greater love, understanding, and achievement in mathematics for the students. In a classroom where active participation is the norm, students can be encouraged to learn from mistakes, use trial-and-error techniques, encounter mathematics in a natural setting, discuss and argue convincingly to persuade other students that their approach to solving the problem is correct, and take responsibility for their own learning. Teachers' roles change from dictator, authority figure, and presenter to those of facilitater, organizer, moderator, and observer. The classrooms themselves change from rooms with desks arranged in five precise rows, chalkboards full of "boardwork" (excessive problems written on the board to keep students busy), and the quietness of the "typical" library where any talking results in a loud "SHHHH!" to rooms where the desks are pushed together in a variety of groupings, activity is a part of each day's lesson instead of repeated, boring "boardwork," and discussions are encouraged (National Research Council, 1989).

One elementary school administration was not happy, as many are not, with their students' scores in mathematics on national achievement tests. As a result, the adopted textbooks were disregarded and the use of manipulatives was emphasized. The administration of the school later reported that the students were doing much better on both state and national tests. Furthermore, when parents called the school after this type of instruction was implemented, they no longer complained; they wanted to get their children enrolled in the program (National Research Council, 1989). The described situation provides evidence that "behind all the compact, abstract symbolism of mathematics there are ideas which must be understood" (Kutz, 1991, p. 135), if children are to experience success in mathematics achievement. Furthermore, the NCTM *Curriculum and Evaluation Standards for School Mathematics* (1989) emphasize the fact that students should be able to "construct number meanings through real-world experiences and the use of physical materials" (p. 38).

MANIPULATIVES AND USES

Good manipulatives are any concrete objects which model any mathematical concept or skill and which students can easily move and rearrange. These materials should help the students formulate mathematical understanding. Manipulatives include counters, sticks, beans, cubes, beads, tiles, pattern blocks, protractors, rulers, geoboards, paper clips, base ten blocks, Cuisenaire rods, tangrams, and any other objects which can be used to foster mathematical understanding. Any inexpensive, safe to handle material can be used including common household items such as poker chips, paper clips, marbles, golf tees, styrofoam packing pieces, and game counters or markers. Regardless of the materials used, teachers should make sure that the items are safe to handle, not too small for easy manipulation, nor conducive to swallowing (Carson & Bostick, 1988). While these materials are more commonly used in elementary classrooms, secondary mathematics teachers should also look for commercially-made materials which are now being used to teach Algebra and other higher level mathematics concepts. Enterprising, creative teachers can also produce their own manipulatives for teaching more advanced skills (Mitchell, 1991).

Any of these materials can be used to provide concrete ways for students to bring meaning to abstract mathematical ideas, learn new concepts, relate new concepts to previously learned ideas, and devise problem-solving strategies. In fact, these many manipulatives should be used for a variety of purposes. Additional purposes include, but are not limited to, fostering creative thinking, actively involving students in mathematical instruction, fostering communication about mathematics, increasing student interest in mathematics, motivating students to learn new mathematics concepts, and accommodating individual learning styles (Mitchell, 1991).

EFFECTIVE USE OF MANIPULATIVES

Learning styles based instruction that incorporates effective use of manipulatives should progress through three stages that are similar to Brunner's idea of enactive, iconic, and symbolic stages of development and learning. In other words, mathematics concept or skill development should progress through three precise stages of development: (1) concrete, (2) representational, and (3) abstract. Developing number sense, the primary goal for all elementary mathematics curricula, can only be developed successfully through the use of concrete objects and/or manipulatives, which is reinforced through pictures, drawings and graphs (representational) and, finally, through oral and written symbols (abstract numbers).

In the case of addition, students should combine sets to arrive at the number of objects in the combined set. However, if the child is not able to conserve numbers, realize that the number of objects remains the same even though they are rearranged, the concept of addition will be impossible to comprehend. Piaget conducted numerous studies with young children and reported that many of them could not conserve numbers. In order for children to develop conservation of numbers, they must explore a number of real objects while learning to count. Through exploration, they will gradually realize that, regardless of the objects used or the rearrangement of a certain number of objects, the number of objects will remain the same. This simply cannot be developed without the use of manipulatives. Once the manipulatives have been used and the teacher observes that the children have an understanding of numbers, then the next stage would be to have students identify the number of items in pictures (representational form). Students should be allowed to pick the appropriate numeral from a stack of number cards or from a grouping of numerals made from sand paper or sponge paper and say the numbers as they count the concrete materials or the pictures of items. One mistake often made in kindergarten programs is that the students spend too much time writing the numerals. The abstract symbolism should be postponed until the students have passed through the concrete and representational stages. As teachers work through the first two stages with the children, they should connect the concept together through three precise steps: (1) model what the students should be doing (counting objects), "model the concept" (Kutz, 1991, p. 138), (2) say the number represented, provide "the name associated with the concept" (p. 138), and (3) show the students the number represented, "the symbol connected with the name" (p. 138–139). Once the students are ready to begin forming the numerals, they should be allowed to complete the task through a variety of materials before they ever engage in pencil-and-paper activities.

Ways to Write Numbers:
Students may have a ball of clay and form the numeral that represents the set of objects; they may "write" the numeral in sand; they may form the numeral out of stiff cookie dough, which the teacher could bake and the children could eat; they could use cake icing to form the numeral on large cookies, and the cookies could be arranged in order and counted before they are eaten; they could use macaroni and glue to form the numerals on construction paper; they can form the numerals on the floor with their bodies, with one, two, or three children working together depending on the numeral being formed.

The amount of possibilities for this type of "writing" of numerals is only limited by the teacher's own creativity, or lack of it. The main point here is to NOT introduce abstract numerals through pencil and paper until after the students have already experienced all of these other activities. The overall goal of mathematics during the early stage of development should be to develop understanding of mathematics through manipulatives. Students should be able to explore and discover relationships in math so that they develop flexibility when dealing with mathematical concepts and ideas (Baretta-Lorton, 1976).

Once the students can conserve number and count to twenty, concrete experiences involving addition can be added to the curriculum. For example:

Addition Problem With Manipulatives:
Here are two crayons. We put four more crayons with them. How many crayons do we have now?

A variety of these problems should be provided through numerous kinds of objects. The children should be guided to pick out two objects, place four more objects with them, and then count to see how many objects are now in the set. Again, this is done before the students ever touch a pencil. As the activities progress, teachers should remember to model use of the manipulatives, say the numbers and show the children the numbers. Pairs of problems should also be added to allow children to begin to experience the commutative principle of addition. For example:

Commutative Principle of Addition:
  A. I have 3 pieces of candy. I put 4 more with them. How many pieces of candy do we have now?
  B. I have 4 pieces of candy. I put 3 more with them. How many pieces of candy do we have now?

Eventually, the children should discover that the rearrangement of the order of the two numbers makes no difference. Defining the property as the commutative property of addition is not necessary at this time. However, this understanding will reduce the number of addition facts that students must learn (Willoughby, 1990).

While providing concrete experiences through manipulatives may be seen by some teachers as difficult and time consuming, teachers should exert the time and effort necessary to provide the students with a wide variety of these experiences in order to make learning mathematical concepts meaningful and interesting. These concrete experiences should be provided any time that a new concept is introduced to a group of students. Teachers must refrain from introducing the abstract symbols too early, because, as stated earlier, this can contribute to problems learning and/or understanding the desired concept. Even instruction concerning algorithms, regardless of grade level, should be initially taught through concrete experiences. Students must have concrete models upon which to base understanding (Kutz, 1991).

Guidelines for Using Manipulatives

Practice Using Manipulatives

The first, and possibly most important, guideline to follow when any teacher is planning to use manipulatives in a mathematics lesson is to practice using the materials before using them with students in the classroom. Teachers must first manipulate the materials, become familiar with the materials, to fully understand how they can and should be used in the classroom. Familiarity with the materials can be accomplished by attending workshops devoted to the use of various or specific manipulatives, taking a math methods class that incorporates effective use of manipulatives, or by watching videotapes that demonstrate effective use of manipulatives and using the materials individually as directed on the videotapes. Two excellent series of videotapes that can be used individually by teachers or in staff development programs to guide teachers when practicing with the materials are produced by Cuisenaire Company of America. One series, *Mathematics: With Manipulatives* (Burns, 1989), is designed for K-6 staff development while *Mathematics: For Middle School* (Burns, 1990) focuses on specific aspects of mathematics instruction in grades 6–8. Throughout each videotape in both series, teachers model effective use of manipulatives, cooperative learning techniques, and effective questioning strategies. Mathematics concepts such as number sense and numeration, whole number operations, geometry, problem solving, statistics and probability, fractions and decimals, patterns and relationships, number systems and number theory, computation and estimation, and measurement are taught through a wide variety of manipulatives and effective teaching strategies.

Free Exploration Time

Giving students initial free exploration time with manipulatives is another guideline for teachers to follow. Although many teachers only view the time spent exploring materials as "play time," children should be able to create, explore, manipulate, and arrange the manipulatives in a variety of ways before direct instruction begins. Regardless of the grade level of the students, the first day or two of mathematics class at the beginning of the school year should be devoted to free exploration of manipulatives which will be used in various lessons throughout the semester. For example, if the teacher plans to use tangrams, base ten blocks, felt fraction pieces, and pentominoes throughout the academic year, the materials could be placed in four different areas in the classroom, students could be divided into four groups and each group could spend ten to fifteen minutes exploring the materials. By the end of the class period, all students should have explored the use of tangrams, base ten blocks, fraction pieces, and pentominoes. Not only will the students benefit from the free exploration, but observant teachers will be able to note various learning style characteristics of the students. Did the students: (1) work alone or in pairs or small groups, (2) arrange or sort the materials in a certain way, (3) discuss what they did with others, (4) sit at a table or desk or sit on the floor, and (5) take any breaks during the time allowed for working with the manipulatives? Furthermore, when students are allowed free exploration during the first few days of class and approximately five initial minutes of each lesson whenever manipulatives will be used, the students will be more likely to perform the directed activities presented by the teacher during the lesson.

Planning Classroom Management Strategies

Planning classroom management strategies is another important guideline to consider when planning to use manipulatives in the classroom. This includes (1) planning behavior guidelines including strategies for distribution and collection of materials, and (2) planning a detailed lesson organized into small specific steps of action which includes questions to encourage higher order thinking skills.

Behavior Guidelines. Teachers are often concerned with using manipulatives in the classroom because they believe that students are easier to manage when using more traditional methods of instruction. However, when behavior guidelines are established, learning style needs are considered, and students are actively involved in learning with the use of manipulatives, behavior problems are often kept to a minimum and persistence to a task may be lengthened. Rules and procedures for classroom behavior should always be established during the first few days of the school year. While these rules should be kept to a minimum, they should address ways to include the needs for informal design, sound when working independently, short breaks, various sociological groupings, mobility, and other learning style elements accommodated in the classroom.

Use of manipulatives in the classroom requires a few additional rules. Students must have established procedures for distributing and collecting manipulatives. Class leaders for the day can handle the responsibility when the materials are housed appropriately. For example, plastic containers are excellent for housing a variety of materials. Plastic gallon ice cream containers could be collected by the students' families and, thus, provide teachers with storage devices at little or no cost. Copy paper or mimeograph paper is often delivered in large cardboard boxes that are also excellent for storage. Teachers could ask the central office or building administration to save the boxes as the paper is used. Manipulatives could be organized by type in the large containers (base ten blocks, pattern blocks, tangrams, etc.), labeled accordingly, and stacked in a corner of the room. Inside the larger containers, plastic zip lock bags could be used to house the necessary number of manipulatives for each group of students. The leaders for the day could retrieve the appropriate labeled container and then give each group of students a plastic bag. As the lesson ends, students would place the manipulatives back in the bags and leaders could collect the bags, place them in the larger containers, and store the materials in the designated area.

Teachers are also often concerned about the noise level in the classroom when manipulatives are used on wooden desks. One way to reduce the noise level is to use 8½ in. by 11 in. sections of felt for a "home base." These precut sections are available for 10¢ to 25¢ per sheet at many discount stores where cloth goods are purchased. If teachers wish to reduce the cost, the felt can be purchased by the yard and cut into sections by the teacher. One sheet is needed for each group of students. When the manipulatives are distributed to the students, the first routine of the day should be to empty the contents of the plastic bags onto the felt pieces ("home base"). Any time the students are not working with the manipulatives, when a concept is being discussed or when only a few of the manipulatives are being used, "home base" provides a centralized location for the materials and also reduces the noise of the plastic or wooden objects being used on student desks or tables (see section of example lesson). These felt pieces can also be used as individual felt boards for students to use in other lessons and are easily distributed, collected, and stored by an additional class leader/helper.

When the manipulatives have been distributed and placed on "home base" by the students, free exploration time has been utilized, and the materials are again placed on "home base," there is another important rule for the teacher to establish and continually enforce.

Classroom Rule:
When you are not directed to use the manipulatives in some way, place your hands in your lap. Whenever you complete the directed task, place your hands in your lap so that I will know who has finished that step of the assignment.

Although the students are told that this procedure helps the teacher see who has completed the directed task, this simple routine also keeps students from rearranging their manipulatives, playing with someone else's manipulatives, and using the manipulatives in other inappropriate ways such as throwing them. If the procedure is continually reinforced and restated during the first few days of direct instruction with the manipulatives, the desired behavior will become an automatic reaction for most students.

Lesson Plans. Planning a detailed lesson organized into small specific steps of action which includes specific questions to encourage higher order thinking skills are not always included in discussions of classroom management. However, the authors of this text believe that a detailed, specific, well thought-out lesson plan helps ensure that the students stay involved in the lesson and, when students are on task, they are less likely to be engaged in disruptive behaviors. If "dead time" occurs as a result of poor planning on the teacher's part, students will lose interest in the lesson and create behavior problems for the teacher. The best way to illustrate these points is to provide an example of a lesson plan that is organized into small detailed steps and contains specific questions. In addition to the manipulatives and felt mats, the following example also includes the use of place value charts. While there are many types of place value charts available for purchase, inexpensive ones are easily made by using laminated sheets of construction paper, dividing the sheet into two, three, or four sections and labeling each section appropriately (ones, tens, hundreds). When a permanent transparency marker is used to divide and label the sections, the writing will not smear as students use the mats and fingernail polish remover can be used to erase/remove the markings when a different type of mat is needed (see Figure 1).

Hundreds	Tens	Ones

Figure 1. Place value charts can be made from laminated construction paper using permanent transparency markers. The charts can be used when teaching place value, addition, subtraction, multiplication, and division.

Lesson Plan For Subtraction With Regrouping
Distribute base ten blocks, place value charts, and felt mats.
Students place base ten blocks on home base.
Students have approximately 5 minutes of free exploration time.
Students arrange their base ten blocks on home base.

T: We already know how to do problems such as this:

(46-24 written vertically on the chalkboard).
      Example: 46
                       − 24
                       ‾‾‾‾

Show me 46 on your place value charts (see
Figure 2).
How many ones do you have on your chart? . . .
(student's name)

Hundreds	Tens	Ones
	‖‖‖‖	▢ ▢ ▢ ▢ ▢ ▢

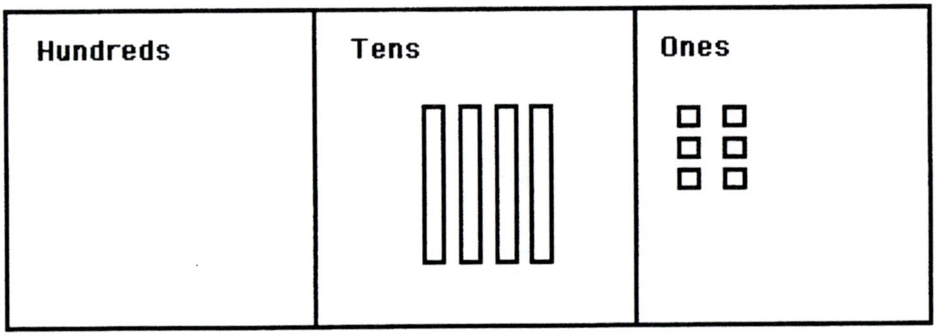

Figure 2. Students demonstrate knowledge of the number 46 by placing 4 tens in the tens' column and 6 ones in the ones' column.

S: 6
T: Why do you have 6 ones? . . . (student's name)
S: Because there are 6 ones in 46.
T: Yes, there are 6 ones in 46 and that is why we have 6 ones on the place value chart. How many tens do you have on your chart? . . . (student's name)
S: 4
T: Why do you have 4 tens? . . . (student's name)
S: Because there are 4 tens in 46.

T: Yes, there are 4 tens in 46 and that is why we have 4 tens on the place value chart. Where do we start subtracting? . . . (student's name)

S: The ones place.

T: Yes, if we are subtracting 24, how many ones do we want to remove? . . . (student's name)

S: 4

T: Can we remove 4 ones? . . . (student's name)

S: Yes

T: Take off 4 ones and place them back on home base. How many ones are left on the place value chart? . . . (student's name)

S: 2

T: Why? . . . (student's name)

S: Because we had 6 and we removed 4 so we have two ones left on the place value chart.

T: Yes. (Record the two on the chalkboard.) Now what do we do? . . . (student's name)

S: Subtract 2 tens.

T: Yes, subtract your two tens. How many tens do we have on our place value charts? . . . (student's name)

S: 2

T: Why? . . . (student's name)

S: Because we had 4 tens, we removed 2 tens and we have 2 tens left.

T: Yes, now let's see what we must do when there are not enough ones to take away. Clear your place value charts and place all of your base ten blocks on home base. Let's consider the subtraction problem 41 − 13. (Write problem vertically on the chalkboard.) Show me 41 on the place value charts. How many ones do you have? . . . (student's name) Why? . . . (student's name) How many tens do you have? . . . (student's name) Why? . . . (student's name) Now we want to subtract 13. Where do we start? . . . (student's name) Yes, we always start with the ones place. Cover your tens with your hand and only look at the ones column. Can we remove three ones? . . . (student's name) No, there is only 1. We are going to have to trade or regroup 1 ten for 10 ones. (Students should have had previous lessons demonstrating that 1 ten is equal to 10 ones and 10 ones are equal to 1 ten and previous practice changing 1 ten for 10 ones and 10 ones for 1 ten.) Trade one ten for ten ones and then stop. How many do we have in the ones column now? . . . (student's name)

S: 11

T: Why? . . . (student's name)

S: Because we traded one ten for ten ones and we already had 1 so now we have 11 ones.

T: Yes. How many do we have in the tens column now? And why? . . . (student's name)

S: 3. Because we removed one ten when we traded and we have one less now.

T: (Record on board by marking through the 4 and 1 and writing 11 in the ones place and 3 in the tens place.) Now can we take away, remove, or subtract 3 ones? . . . (student's name) Yes, do it. Now how many ones are left on the chart? . . . (student's name)

S: 8

T: Why? . . . (student's name)

S: Because we had 11 ones and we removed 3.

T: Where do we subtract now? . . . (student's name)

S: The tens column.

T: We want to take one ten away. Do this. How many tens do we have left? . . . (student's name)

S: 2

T: Why? . . . (student's name)

S: Because we had 3 tens after we traded and we took one away and now we have 2 left.

T: So if we take 13 from 41 we have 2 tens and 8 ones left. What is the answer to our problem? . . . (student's name)

S: 28

T: Clear your place value charts. Place your blocks back on home base. Then do 53 − 17 and other problems from the textbook in the same way.

Please note that nowhere in this introductory lesson on subtraction with regrouping were the students asked to write or record anything. The teacher lead the students through detailed steps as she explained and recorded what the students were doing on the board. She modeled the correct behaviors and processes while she ensured that the students understood the concept of subtraction with regrouping. Through teaching subtraction with regrouping in this way, students understand why certain numbers have lines through them and others are written in their place. If they look at one place at a time (ones), they understand when regrouping is needed and when it is not. This will also reduce problems of subtracting smaller numbers from larger numbers even when the top number is smaller, a mistake students often make.

Specific questions are also included to encourage students to consider and explain orally why each step was completed. Students must not only routinely complete steps of a process as directed by the teacher, but must also understand and explain why each step is completed in order for mastery of the concept to be reached. When teachers do not take the time to plan the specific steps used in working a specific type of problem and questions to ask to foster understanding of why each step is completed, necessary parts of the process are often overlooked as the teacher presents the lesson in the classroom and thought questions are seldom asked. As in the previous example, detailed plans for working one problem can serve as a guide for similar problems and restating the same steps and questions for each problem worked during the lesson is not necessary. However, teachers should use problems similar to the ones presented in the textbook and the problems they wish to use should be recorded so that composing problems from memory during the lesson (which could lead to the use of inappropriate problems) will not be necessary.

Once students master using the manipulatives as included in the above lesson, the next step in the process is to have the students record each step of the problem using pencil and paper as the manipulatives are used. Watching the teacher record the steps on the board as manipulatives are used should provide an easy transition. After guided practice using the manipulatives and recording each step as directed by the teacher, then students can work a complete problem with the manipulatives and record the process before the problem is discussed and recorded on the board by the teacher. Before requiring students to work the problem with only pencil and paper, each phase for using manipulatives should be completed.

Phases Of Manipulative Use:
1. Manipulatives only
2. Manipulatives with pencil and paper for each step, and
3. Working the complete problem with manipulatives and pencil and paper.

Furthermore, students should be allowed to use the manipulatives until they feel confident completing similar problems without their use. Most students will stop using the manipulatives as the concept is understood; some may wish to have the manipulatives on their desks and only use them for more difficult problems. This "weaning" of the manipulatives should not be rushed by the teacher. When teacher tests for understanding are administered, students should still be allowed to use the manipulatives. Students can be encouraged to try the problem without the manipulatives but should be allowed to use the manipulatives until the concept is mastered. However, since manipulatives are not used with many standardized tests, the manipulatives will have to be eliminated before such tests are given to students.

ACCOMMODATING LEARNING STYLE NEEDS WHILE USING MANIPULATIVES

Learning style profiles of students should be considered by teachers when planning for use of manipulatives. However, needs of individual students are often overlooked when these materials are used. Students usually work in groups of two to three when manipulatives are used because of the number of manipulatives needed, the cost of manipulatives, and the advantages of students working in groups. When students work in pairs or small groups, they are provided with natural situations in which they can talk about mathematical ideas, listen and evaluate the ideas of others, learn through making mistakes in a nonthreatening environment, gain confidence in their own mathematical abilities (Kutz, 1991), and develop and strengthen abilities to form mathematical concepts and solve problems (Cangelosi, 1988). In small group activities using manipulatives, each child functions as a teacher as well as a learner. When the child is acting as the teacher, s/he must clarify mathematical ideas in his/her own mind to be able to explain the activity to other students, which improves understanding of the language and the mathematical skills involved in the lesson (Kutz, 1991). Furthermore, the learning style profile of most underachieving or poorly achieving students indicates that they are typically anti-adult, which means they do not learn well in one-on-one remedial situations with teachers or teaching assistants (Dunn, 1988; Midkiff, 1991; Midkiff & Towery, 1991a; Midkiff & Towery, 1991b; Midkiff, Towery, & Roark, 1991a; Midkiff, Towery, & Roark, 1991b). Using manipulatives in small groups is especially effective for these students, since they are working with their peers, hearing and seeing the explanations of their peers as they work with the manipulatives, and verbalizing their own understandings in an environment that fits their individual learning styles. Often these students can handle difficult problems or situations in small groups that they would not be able to master in isolation. However, this type of learning environment will not meet the learning style needs of students who prefer to work alone. If students have a high need to work alone and the required number of materials is available, these students should be allowed to work independently in either a formal or informal design as indicated by the students' learning style profile.

In fact, flexible teachers can accommodate a variety of learning style needs when manipulatives are used. For example, teachers can form groups of students who all prefer an informal design and allow these students to sit on the floor while using the materials. Students who prefer a more formal design can be placed around tables or can arrange their desks in groups. The need to work in either a brightly or dimly lighted room can be accommodated in a similar way. Obviously, the need for sound is not usually met when students work in groups, but if a student is working alone, Walkmans® can be used to provide sound when s/he reaches the stage of working complete problems with or without the manipulatives during independent practice.

When using manipulatives in the classroom, teachers should expect the right-brained, global learners (who may be underachievers in mathematics since it is typically taught in a more left-brained analytic way) to initially do better than other more analytic learners. When one of the authors of this book taught a sixth grade lesson on long division, the students who were the higher achievers in the mathematics classroom had trouble translating the abstract steps involved in long division to use of base ten blocks. They had memorized the steps, but did not understand the process. The "underachievers" in the classroom, however, quickly understood how to manipulate the blocks and record each step. Once these students grasped the concept, they explained it to the "smarter" students who eventually gained understanding of the meaning of each recorded step. Not only was understanding increased in all of the students, but the self-esteem of these "underachievers" was enhanced when they had the opportunity to explain mathematics to the students who usually earned higher grades in this mathematics classroom.

CONCLUSION

Classrooms that incorporate use of manipulatives, concrete objects, real-life experiences, and active learning, while accommodating learning style needs of students, require a great deal of time and effort on the part of teachers. However, the results, greater love, understanding, and achievement in mathematics for the students, will be well worth the time and effort involved. Furthermore, this type of mathematics instruction enables students to have a good foundation in mathematics, which is required before higher level concepts and skills can be learned. Any time manipulatives or concrete objects are used, teachers should model the appropriate behavior, discuss and explain the concept or skill, and record the abstract version for the students. Additionally, students should be allowed to move gradually from using the concrete materials to the representational stage (pictures) before reaching the abstract level of mathematics operations (working the problems with pencil and paper).

Careful preparation and planning by the teachers are partially responsible for the time involved in using the constructivist method of mathematics instruction. Guidelines for teachers to follow when planning and using manipulatives include becoming familiar with the manipulatives before using them in a lesson, allowing free exploration time before directed activities are presented to the students, and planning classroom management strategies which include a detailed lesson plan with questions to elicit higher order thinking and reasoning from students. A simple technique, having students place their hands in their laps as they complete each directed step, will allow teachers to monitor completion of each assigned task and will also greatly reduce behavior problems commonly associated with using manipulatives in the classrooms.

Although students are usually placed in groups in formal design areas when using manipulatives, teachers should consider learning style needs of students when planning for use of these materials. Sociological grouping preferences, formal and informal design needs, lighting preferences, and other elements of learning style are easily accommodated in the classroom while manipulatives are used to provide basic understanding of the mathematical concept or skill being taught. When learning style needs are accommodated as manipulatives are used, an excellent foundation in mathematical understanding will be developed in the students.

Chapter Five

DIMINISHING GENDER DIFFERENCES IN MATHEMATICS ACHIEVEMENT

In elementary school, similarities between the sexes in mathematical achievement have been more frequent than differences, and any difference has typically favored the females. In junior high school or middle school, approximately half of the few sex differences in mathematics have also favored the females. However, by the time students reach high school and college, any sex differences have favored the males (Cambell, 1986). Franklin (1990b) reported that females entering elementary school have appeared to enjoy mathematics and have done well in mathematics achievement, but as these students progress through the elementary years, this interest in mathematics has tended to decrease. At each grade level from six to eleven, boys have appeared more confident of their mathematical abilities (Fennema & Sherman, 1987). Throughout history, differences in achievement have tended to increase with age (Benbow, 1986; Fennema, 1974). Furthermore, this has been most noticeable among the gifted students (Benbow, 1987; Maccoby & Jacklin, 1974).

Much research and many publications have addressed the issue of gender preference in the area of achievement in mathematics. Noted researchers such as Aiken (1975), Armstrong (1975), Benbow & Minor (1986), Fennema (1975), Fox (1981), Fruchter (1954), Hilton & Berglund (1974), Maccoby & Jacklin (1974), Meeker (1979), Mitchell & Burton (1984), Stage & Kurplus (1981), and Tobias (1976, 1978) have isolated several probable reasons for the discrepancy in mathematics achievement. These reasons include physiological differences, societal expectations, toys and games typically chosen by males and females, and spatial perception skills.

PHYSIOLOGICAL DIFFERENCES

Sex differences in mathematical performance are biological, according to Benbow (1986, 1987). She contended that a high percentage of fetuses who had been exposed to high levels of testosterone were left-handed and prone to allergies, traits twice as common in mathematically-gifted students as in other students. Medically, testosterone has been found to slow the development of the left hemisphere of the brain, which creates a situation in which the right hemisphere's development was accelerated. This may be desirable since the right hemisphere of the brain is the center for the reasoning ability needed in mathematical problem solving.

Authorities have also contended that sex differences occur in the way the brains of males and females deal with certain mental tasks (Willis, Wheatley, & Mitchell, 1978). High school students who were tested in an attempt to determine if there were differences in the learning styles and brain patterns of males and females were divided into two groups to perform analytical/linguistic tasks and spatial tasks. Females and males both tended to use the left hemisphere to process analytical/linguistic tasks. However, different hemispheres of the brain were used by females and males when processing spatial tasks. The females tended to favor the left hemisphere, while the males tended to favor the right hemisphere. By the time the two gender groups were in high school, both hemispheres of the males' brains had been exposed to exercise and the females' brains had received exercise in only the left hemisphere.

SOCIETAL EXPECTATIONS

Authorities (Fennema & Sherman, 1987; Fox, 1975; Hilton & Berglund, 1974; Tobias, 1978) who have conducted research during the last several decades have attributed the differences in mathematical performance to the theories that the social climate favored males in mathematical achievement and that females received different treatment from peers, teachers, and parents. Fox (1975) found that females were less likely to seek out special experiences related to mathematics since they tended to have values and interests of a more social nature. Many times, getting talented females and their parents to agree to participate in an accelerated mathematics class has proved difficult due to a fear of social ostracism (Fox, 1981). American society has tended to make mathematical ability a masculine attribute and has punished women for doing well in mathematics (Tobias, 1976). Currently, an emphasis has been placed on the importance of equity in mathematical performance for females. The Women's Equity Act Program under the auspices of the United States Department of Education is one of many agencies striving to provide awareness and remediation activities for females in mathematics.

EFFECTS OF TOYS AND GAMES IN ACHIEVEMENT OF MATHEMATICAL SKILLS

In a research study conducted by Pattison and Grieve (1984), any sex differences found as a result of mathematical testing/spatial perception were in favor of the males. Tobias (1976) theorized that this could be due to the toys young males often choose, including building blocks and take-apart toys which are often more accessible by males than females. Furthermore, math readiness was found to be enhanced by games and toys that lead to an understanding of shapes and how things work (Grayson & Martin, 1988). Studies have consistently revealed that females of all ages tend to do more poorly than males in spatial perception relations (Fennema & Sherman, 1977; Young, 1982). According to Tobias (1978), comparing males and females of all levels of intelligence has continued to reveal that males, on the average, performed better than females when completing various spatial perception activities such as reading maps, finding embedded hidden figures, learning mazes, and solving geometric problems.

SPATIAL PERCEPTION SKILLS

Students displaying well-developed spatial intelligence tend to think in patterns or images and excel in reading maps and charts. They may also be quite adept when manipulating puzzles and images, and frequently enjoy building three dimensional objects or inventing projects of a mathematical nature (Armstrong, 1987). Visual thinking skills allow students to mentally rotate objects and designs which results in high performance in geometry and trigonometry. According to Blackwell (1982), there are two main factors in spatial perception: (1) spatial visualization and (2) spatial orientation. Both of these abilities reflect one's relationship to the view of a given object. Furthermore, spatial perception skills usually involve two or three dimensional objects and, possibly, their rotation (Tobias, 1976). Well-developed spatial skills allow students to recognize relationships of figural analogies and similarities necessary for higher levels of mathematics.

Even though there has been much controversy over the many different theories concerning the difference in the mathematical performances of males and females, a distinct correlation appears to exist between spatial perception skills and performance in mathematics (Fennema & Sherman, 1977; Mitchell & Burton, 1984; Stage, Kreinberg, Eccles, & Becker, 1985). If the mathematical problem to be solved includes spatial or visual clues, the student with high spatial perception skills will learn more than the student demonstrating low spatial perception skills (Battista, 1981). A strong relationship has also been found between mathematics achievement and spatial perception skills requiring students to select three-dimensional shapes that are formed by folding two-dimensional shapes (Chipman & Wilson, 1985).

From an early age, males have been encouraged to participate in activities which develop spatial perception performance, including playing with blocks and toys that require large muscle efforts (Willis, Wheatley, & Mitchell, 1978). Females, on the other hand, have tended to read earlier than males. "Girls, except those who had manipulative materials in Montessori school and/or have been actively involved in sports before age 5, have very poor figural [spatial and often right brain function] abilities" (Meeker, 1979, p. 301). Conversely, males have typically been more engaged in activities such as strategy-memory contests (chess) and playing "geometrical or trigonometrical sports (sailing and billards)" (Tomizuka & Tobias, 1981, p. 114) outside of school than have females. However, spatial perception abilities of students typically mature between the ages of 11 and 15 (Fruchter, 1954) and the use of spatial foundations for mathematics becomes necessary. During this age, females have typically been less able to understand spatial perception relationships than males (Meeker, 1979).

CURRICULUM AND SPATIAL REASONING

Since experiences designed to nurture spatial perception skill development are usually not taught in school (Stage, Kreinbert, Eccles, & Becker, 1985), students deficient in these skills often have difficulty in acquiring them. If the male dominance in the mathematical field is true, educators need to evaluate the methods used in the teaching of mathematics to females. Alternative methods of mathematical instruction could be used in an effort to improve females' mathematical performance (Buerk, 1985; Weiner & Robinson, 1986). In the former Soviet Union, mathematics instruction has been changed to incorporate learning experiences involving spatial perception skills (Young, 1982). The traditional American viewpoint has been that one's individual spatial perception abilities are unchangeable; however, recent researchers believe that experiences with mazes, reading maps and diagrams, rotating dice, inverse drawings, soma cubes, Origami, tangrams and geoboards affect these abilities (Armstrong, 1987; McKim, 1980; Mitchell & Burton, 1984; Stage, Kreinbert, Eccles, & Becker, 1985; Tobias, 1978).

Reasoning skills, including spatial reasoning, are required for students to be able to know, understand, and work successfully with mathematics. In fact, the *NCTM Standards* (1989) include Mathematics as Reasoning as one of the standards at all levels of mathematics achievement, K–4, 5–8 and 9–12. Reasoning should be an integral part of all mathematics to develop deductive and inductive reasoning, spatial reasoning, and reasoning with proportions and graphs. However, special attention should be given to the development of spatial reasoning to attempt to reduce the gender differences often found in mathematical achievement.

Strategies involving spatial concepts such as three-dimensional space and shapes should be used in teaching mathematical concepts rather than ditto sheets and workbook pages (Cherry, Goodwin, & Staples, 1989). "Especially in the early elementary years, make sure that the girls . . . play with blocks, Legos, and other manipulative building toys as much as the boys do (Franklin, 1990a, p. 225). Through specific activities which involve spatial reasoning, the difference in spatial perception found in males and females can be significantly reduced (Blackwell, 1982).

Preschool students begin to develop an understanding of mathematical concepts through the use of manipulatives including, but not limited to, blocks, puzzles, cubes, and construction toys. These activities have also proven beneficial for primary students who are experiencing difficulty in understanding the concepts of math (Thomasson, 1988). The replacement of skill sheets and paper-and-pencil activities with manipulatives gives the left dominant students experience in developing the spatial awareness while the right dominant students are learning through their preferred modality (Cherry, Godwin, & Staples, 1989).

SPATIAL REASONING SKILLS

When spatial reasoning skills including visual thinking skills (mental operations that are sometimes assessed in intelligence tests) are developed, greater success in higher levels of mathematics is often achieved (McKim, 1980). Examples of visual thinking skills activities include memory games involving shapes and spatial rotation, as well as rotation of one-two-and/or three-dimensional figures, hidden shapes, matching and cut out forms.

Categories of Visual Thinking

Memory of Shape

Memory of shape activities require students to view a shape and recall its configuration.

Memory of Shape Activity:
The student will closely observe a shape or a picture. The object is removed, and the student attempts to draw the picture in detail. If unable to complete the task, the student may view the object a second time. Suggestions for increasing the student's ability to recall the same picture include tracing the outline with the finger and focusing on specific perimeters of the shape. For students experiencing extreme difficulty in attaining this skill, reduce the number of shapes to be reproduced. After the student can reproduce two or three objects, increase the number of objects to four or five, etc. until harder problems can be completed (see Figure 3).

Pattern or Figure Completion

Pattern or figure completion activities involve mentally completing a figure's or object's lines in order to identify the object or complete the picture.

Pattern or Figure Completion Activity 1:
Jigsaw puzzles are excellent manipulatives for pattern and figure completion. The student should progress from simple puzzles with few pieces to the more difficult puzzles which might have several pieces of the same shape but different picture designs.

An excellent learning center for an elementary classroom is a puzzle center. Students can collectively complete the multi-pieced puzzle over a period of several days.

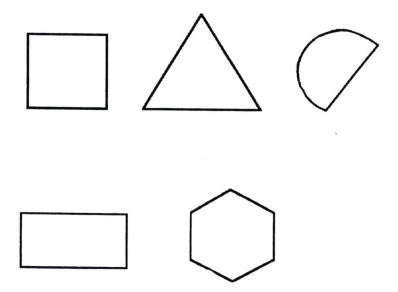

Figure 3. Students reproduce geometric shapes.

Pattern or Figure Completion Activity 2:
The teacher can utilize nature as a figure completion activity. Moving the classroom to the outdoors will allow the students the opportunity to observe the cloud formations and visualize shapes and objects created by the shapes of the clouds.

One way to facilitate mathematical connections as advocated by the *NCTM Standards* (1989) would be to incorporate this activity into a science unit on clouds and have the students write about the patterns seen in the cloud formation which also involves mathematical communication.

Pattern or Figure Completion Activity 3:
A higher degree of pattern or figure completion skills involve a paper and pencil activity of completing the missing segment. The experience of manipulating the jigsaw pieces provides important background experiences to the one-dimensional tasks involved with pencil and paper (see Figure 4).

For the students experiencing extreme difficulty in completing the sequence of this type, use a manipulative approach. Have the students cut a section of the beginning pattern and place it underneath the paper under the open space. Trace the recurring pattern onto the blank paper. The students should be allowed to trace several problems until

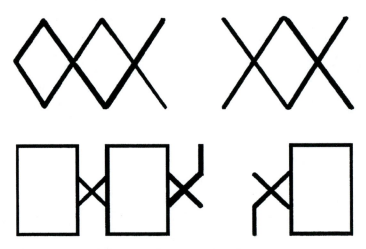

Figure 4. Students complete the pattern.

they can successfully complete a problem without the assistance of tracing. An extension activity could involve having students create figure/pattern completion puzzles for other students to solve.

Pattern or Figure Completion Activity 4:
The student completes a design to create an artistic picture. The patterns of the geometrical shapes are colored so as to create pictures or mosaics.

Additional pattern or figure completion activities are available in:

Bezuska, S., Kenney, M. & Silvey, L. (1977). *Tessellations: The geometry of patterns.* Palo Alto: Creative Publications; and

Finkel, L. G. (1980). *Kaleidoscopic designs and how to create them.* New York: Dover Publications.

Rotation of One-Dimensional Figures

With rotation of one-dimensional figures, the student can perceive how a pattern on a paper would appear if it were rotated. The activities presented to develop this visual thinking skill are presented in hierarchial order. The first activities involve skills that are the easiest to acquire. Scissors and dot or graph paper allow the student to transform the paper and pencil tasks to a manipulative one. Once the rotation skill is acquired through manipulatives, then the student should progress to the paper and pencil format. If the students experience difficulty in completing the paper and pencil tasks, they should be allowed to return to the manipulative approach for additional help.

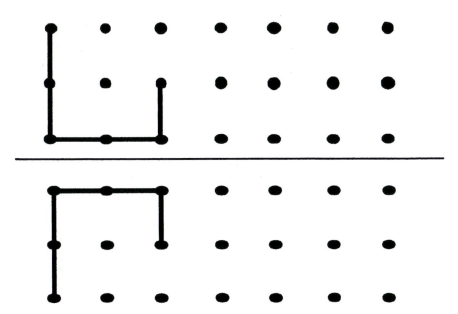

Figure 5. Students produce the mirror image of the top pattern.

Rotation of One-Dimensional Figures Activity 1:
The drawing in the box is drawn onto the dotted paper in a reversed pattern. See Figure 5.

Additional activities of this type can be found in:

Lund, C. (1980). *Dot paper geometry with or without a geoboard.* New Rochelle: Cuisenaire Company of America.

Rotation of One-Dimensional Figures Activity 2:
The series of drawings are rotated to form a pattern. The student is asked to actually darken the portion of the figure which completes the pattern (see Figure 6).

Rotation of One-Dimensional Figures Activity 3:
The first figure presented is shown from another viewpoint. The student is asked to identify the same figure in a different position. One clue to aid the student is to trace the various shapes in the first picture. The same shapes should be present in some form in the correct answer (see Figure 7).

Additional activities of this type are available in:

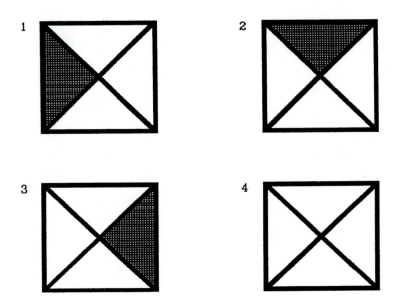

Figure 6. Students complete rotation pattern by darkening the appropriate section.

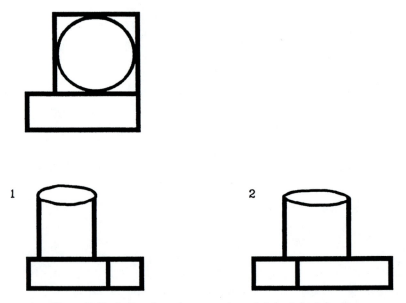

Figure 7. Students select alternate view of the original drawing.

Behavioral Research Division, Institute for Applied Research Services—University of New Mexico. (1982). *Spatial encounters: Exercise in spatial awareness.* Newton: WEEA Publishing Center.

Rotation of One-Dimensional Figures Activity 4:
Line design pictures are created by connecting points on a circle with straight lines. This operation is done with a compass and a straight edge. Different colored pens and pencils create a pleasing design (see Figure 8).

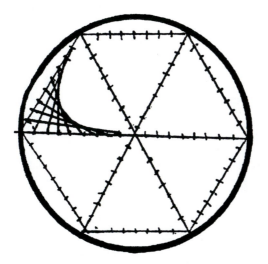

Figure 8. Students produce designs by connecting points on a circle with straight lines.

Additional activities are available in:

Seymore, D., Silvey, L., & Snider, J. (1974). *Line designs*. Palo Alto: Creative Publications.

Rotation of One-Dimensional Figures Activity 5:
The student is asked to draw the last figure in the analogy sequence. The dotted square aids the student in producing the final product. The first figure could be cut out to make a manipulative to rotate for the following figures (see Figure 9).

Additional activities are available in:

Black, H. & Black, S. (1981a) *Figural analogies A-1.* Pacific Grove: Midwest Publications.
Black, H. & Black, S. (1981b) *Figural analogies B-1.* Pacific Grove: Midwest Publications.
Black, H. & Black, S. (1981c) *Figural analogies C-1.* Pacific Grove: Midwest Publications.

Rotation of One-Dimensional Figures Activity 6:
The student is asked to complete the figure by the addition of other pieces or shapes. The student could trace the size of the empty space or use a ruler to measure the size of the space. The possible choices could be cut out and used as manipulatives to actually fill the space.

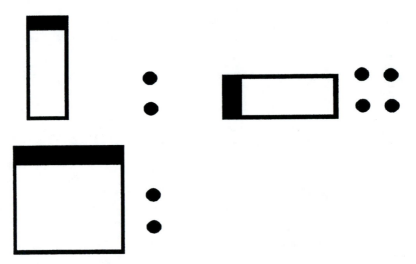

Figure 9. Students complete geometry analogy by drawing the fourth number of the set. The semicolon represents the words, "is to" and the double semicolon represents the word, "as."

This activity lends itself quite easily to moving from the concrete stage to the representational phase (see Figure 10).

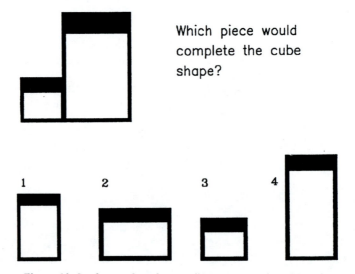

Figure 10. Students select the one figure to complete this cube.

Additional activities are available in:

Black, H. & Black, S. (1983a) *Figural similarities A-1.* Pacific Grove: Midwest Publications.
Black, H. & Black, S. (1983b) *Figural similarities B-1.* Pacific Grove: Midwest Publications.

Black, H. & Black, S. (1983c) *Figural similarities C-1.* Pacific Grove: Midwest Publications.
Black, H. & Black, S. (1983d) *Figural similarities D-1.* Pacific Grove: Midwest Publications.

Rotation of One-Dimensional Figures Activity 7:

A design is drawn on the paper and the student is asked to determine what the design would be when it was folded. The students can actually cut out the design to make it a two or three dimensional figure. After practicing, the students might want to take a paper figure that has been folded and unfold it to see the initial pattern (see Figure 11).

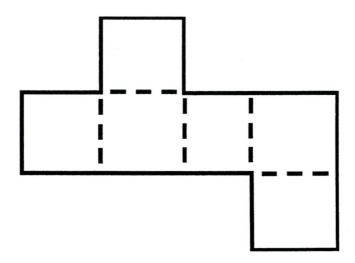

Figure 11. Students determine what three-dimensional shape will be produced by folding the two-dimensional drawing.

Rotation of One-Dimensional Figures Activity 8:

The students are asked to either determine what a folded paper would look like when unfolded or what an unfolded paper would look like when folded. The method to use in aiding the students in this task is to allow them to actually fold or unfold the paper. In the example in Figure 12, the paper is hole-punched, and the position of the holes is located. The students should use the paper and hole-punch to solve the problems until they feel secure in solving the problems without the aid of manipulatives (see Figure 12).

Additional activities are available in:

Hornadek, A. (1979). *Spatial perception—A.* Pacific Grove: Midwest Publications.

When the figure is folded on the dotted line, where will the holes be in the other section?

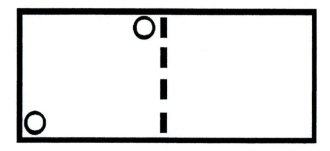

Figure 12. Students determine mirror image placement of the holes.

Rotation of Two and/or Three Dimensional Figures

Rotation of two and/or three dimensional figures results in rotation of a manipulative in some designated manner. The manipulatives can be paper that will be folded, cubes that will be moved, or string that will be knotted. The end product is the result of the manipulative being rotated in some manner.

Rotation of Two and/or Three Dimensional Figures Activity 1:
The student either reads directions or follows diagrams to produce knots. The manipulative can be rope, twine, string, lightweight wire, or any items that can produce the same results. *The Boy Scout Manual* and *The Girl Scout Manual* give excellent instructions in this task (see Figure 13).

Rotation of Two and/or Three Dimensional Figures Activity 2:
Folded paper items range from a simple folded hat to involved folded animals. The art of paper folding to produce designs is called Origami and is quite prevalent in the Oriental cultures. The directions are usually given in written and diagram form.

Activities of this type are available from:

Silvey, L. & Taylor, L. (1976). *Paper and scissors polygons.* Palo Alto: Creative Publications.

Stangl, J. (1984). *Paper stories.* Belmont: Fearon Teacher Aids.

Students tend to like this type of activity, and since there are many levels of difficulty, the students can usually find a specific activity at the appropriate level of difficulty.

Figure 13. Students will tie a knot by following written directions.

Rotation of Two and/or Three Dimensional Figures Activity 3:
Tangrams and pentominoes enable students to create a design with the shaped pieces or completely fill a space with the shaped pieces. This activity allows for creativity as well as space conservation.

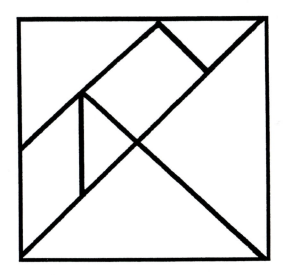

Figure 14. Students arrange the 7 Tangram pieces to form a square.

Tangrams are ancient Chinese puzzles that are still used today by students of all ages for recreation as well as educational purposes. A tangram begins with a square which is then cut into seven pieces: two large triangles, one medium triangle, two small triangles, a square and a parallelogram (see Figure 14). These puzzle pieces can be used to form various geometric shapes in addition to the square. The seven pieces can also form a rectangle, parallelogram, trapezoid, and triangle, among others. Each of these shapes including the square can also be made using the two small triangles and the medium triangle or can be made using the five smaller pieces. In addition to teaching geometric shapes and spatial reasoning skills, these manipulatives are excellent to use when teaching geometric shapes, fractions, area, and perimeter. When students work in pairs or small groups to manipulate the tangrams, their mathematical communications (*NCTM Standard:* Mathematics as Communications) skills are developed and increased understanding of the concepts being studied are developed.

Grandfather Tang's Story: A Tale Told With Tangrams (Tompert, 1990) is an excellent resource for teachers using tangrams in the classroom. The tangrams are arranged in a variety of ways to depict various animals discussed in the story. Students could act as story tellers and use the tangrams to show the shape of each character or element in their story. As a new character is introduced, the shapes will be rearranged to denote the new character.

Pentominoes are geometric shapes that are formed by joining five squares edge-to-edge. The name is derived from "domino," the shape made by joining two squares edge-to-edge. Twelve different shapes can be made by placing five squares edge-to-edge. The twelve pentominoes can be placed together in a variety of ways to form a rectangle, or smaller groups of the pentominoes can form a variety of different shapes. Figure 15 shows nine pentominoes formed to make an L-shape.

Additional activities of this type are available from:

Picciotto, H. (1984). *Pentomino activities lessons and puzzles.* Sunnyvale: Creative Publications.

Rotation of Two and/or Three Dimensional Figures Activity 4:

Pattern blocks are used for rotation on a given axis. Students can complete the design on the paper and then rotate or mirror image the design with the use of the manipulative pieces (see Figure 16). After completing the provided sheets, the students could create problems for others in the class to solve.

Pattern blocks are collections of green triangles, yellow hexagons, red trapezoids, blue parallelograms, white rhombi, and orange squares. In addition to being used for spatial reasoning and problem solving skills, they are also excellent to use when teaching geometric shapes, fractions, area, perimeter, and other mathematical concepts.

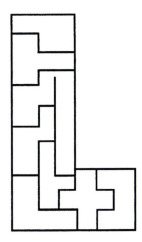

Figure 15. Students use 7 Pentomino pieces to fill the L-shape.

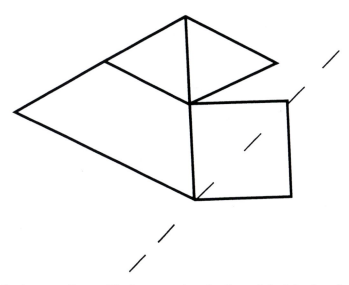

Figure 16. Students use Pattern Blocks to complete the figure left of the dotted axis and then build its mirror image.

Additional activities for using pattern blocks are available from:

Pasternack, M. & Silvey, L. (1975). *Pattern blocks activities B.* Palo Alto: Creative Publications.

Rotation of Two and/or Three Dimensional Figures Activity 5:
Matches are inexpensive items which can be an excellent source of design completion and puzzle solving. The matches are placed in a

distinct pattern to begin, and the students create another pattern by moving or removing certain matches (see Figure 17).

Remove 5 matches to leave 3 squares.

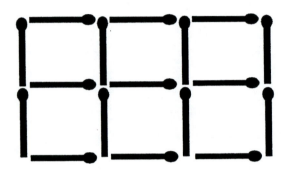

Figure 17. Students use spatial reasoning and problem solving skills to solve the match problem.

Rotation of Two and/or Three Dimensional Figures Activity 6:
The student is asked to determine which face of a die would be upward when the die is rotated. The activity lends itself to the manipulative approach so that the student can actually rotate the die to determine the answer. After experience with the manipulative, the student should progress to the representational and abstract stages (see Figure 18).

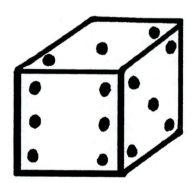

Figure 18. Students determine which face of the die would be upward when the die is rotated.

Rotation of Two and/or Three Dimensional Figures Activity 7:
Soma cubes lend themselves to a multi-dimensional approach. There are many cubes that are connected to form a design. The student is asked to determine which design will be formed if the original design is rotated. The student can use multi-link cubes to actually make the manipulative to rotate and after practice should strive to move to the representational and abstract stages of locating the rotated design on paper (see Figure 19).

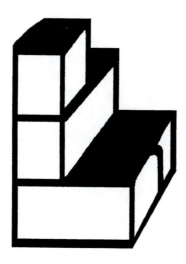

Figure 19. A variety of three-dimensional objects are produced by combining cubes.

Rotation of Two and/or Three Dimensional Figures Activity 8:
Tinkertoys, building blocks, legos, electrosets, and capsela are all building types of toys found in toy departments of local discount stores or various catalogs. These manipulatives provide excellent experience for various spatial perception skills. The students have the ability to rotate and build configurations during free exploration time or in planned class activities.

ACCOMMODATING LEARNING STYLE NEEDS WHILE USING SPATIAL REASONING ACTIVITIES

Activities which develop spatial reasoning skills can easily be incorporated into the elementary mathematics curriculum. If the activity involves working with a one-dimensional activity such as paper and pencil rotations, the student might also benefit from creating a manipulative out of the paper. The tactual or kinesthetic learner will benefit from the experience; however, the manipulatives will also create a concrete experience for the visual and auditory learners (see Chapter 4). If the students work in pairs or small groups, then the auditory learner can also be accommodated through the explanations and discussions of the group.

Learning style preferences other than modality strengths are also easily accommodated when using spatial reasoning or visual thinking activities. Students can be allowed to sit in an informal or formal design and work in a variety of sociological groupings. Most often, students are encouraged to work in small groups or pairs when completing these activities, but if a student has a strong desire to work independently and can complete the simple activities alone, this student should be allowed to work in this manner. If this student's learning style profile includes the need for sound, Walkmans® can be used when large group instruction by the teacher is not being used. Choices in meeting the individual learning style profiles of the students enables the flexible teacher to ensure optimal learning for all students and may increase motivation, persistence, and responsibility of the students (see Chapter 3).

CONCLUSION

Proportional and spatial activities are needed to develop students' abilities to reason mathematically regardless of the gender or the learning style preferences of the students. Furthermore, development of spatial reasoning abilities may assist in reducing the discrepancies of mathematical achievement often noted between males and females. Concrete objects are often needed to develop spatial reasoning, regardless of the modality strength of the students. The activities described through this writing are only a sampling of ideas on which teachers may build. Teachers may also wish to investigate the use of fractals as a way to enhance spatial reasoning of students. A variety of publications are available from the National Council of Teachers of Mathematics which address this topic.

Chapter Six

MATCHING ACTIVITIES AND LEARNING STYLES

The activities presented in this chapter are organized by learning style characteristics with the primary perceptual preference listed first. Most of them are intended for use as practice activities to reinforce specific skills after the concepts and skills have been reinforced through the use of concrete materials following the guidelines presented in Chapter 4. Drill and practice is still a necessary part of instruction. However, drill and practice sessions should be short, varied in format, and appropriate for the learning style characteristics of the students. The skills reinforced through each activity are listed for easy reference by the reader.

Many of these activities were created by teachers in the Blytheville, Arkansas school district, after the general type of activity was introduced to them. Their willingness to have these activities included in this writing is appreciated and the names of the teachers and the grade levels they teach have been included. Occasionally, the activities have been adapted or edited by the authors of this text for increased understanding by the reader.

AUDITORY, SMALL OR LARGE GROUP ACTIVITIES

MentComps

MentComps are mental computation games especially appropriate for the auditory learners. Any skill or concept can be reinforced through the use of this type of activity. The teacher simply makes a set of cards with an answer to a question on the top half of the card and a different question on the bottom part of the card. The top of the next card has the answer to the previous question and another question is printed on the bottom of that card. This process is repeated until the desired number of cards is produced. On the last card the question on the bottom will be answered on the top of the first card created by the teacher (see Figure 20). In this manner, a circle of questions and answers is produced. Any number of cards can be produced by the teacher, depending on whether the teacher wishes to make this a small group or large group activity. The game is then played by distributing *all* of the cards to the students. Any identified student begins by reading his/her question. The student with the appropriate answer reads his/her card and the process continues until the first reader states the answer on his/her card. Cut out shapes or 3 × 5 inch note cards can be used to make a MentComp activity. Again, each card should have the answer to the previous card's question, followed by a new question that will be answered by the next card. Each answer number should be used only once. In this manner, every student will have to correctly answer a question, then ask a new one for somebody holding the correct card to answer. This can be applied to any subject matter. Care should be taken so that no cards are lost or the chain of questions and answers will be broken. Laminating the individual cards will increase the life-span of the game. This is an excellent activity for a review. Each student will receive one card (be sure there are enough, some can have two if there are extras).

Specific MentComp Activities

Addition and Subtraction MentComp Activity:
Skills: Basic addition and subtraction facts
Each of the following lines is printed on an individual card. Comic figures or cartoon characters can be added for visual interest.
My number is 17. Who has my number minus 8?
My number is 9. Who has my number plus 3?
My number is 12. Who has my number minus 6?
My number is 6. Who has my number minus 5?
My number is 1. Who has my number plus 3?
My number is 4. Who has my number plus 7?
My number is 11. Who has my number minus 9?
My number is 2. Who has my number plus 6?
My number is 8. Who has my number plus 8?
My number is 16. Who has my number minus 9?
My number is 7. Who has my number minus 2?
My number is 5. Who has my number plus 5?
My number is 10. Who has my number plus 3?
My number is 13. Who has my number plus 1?

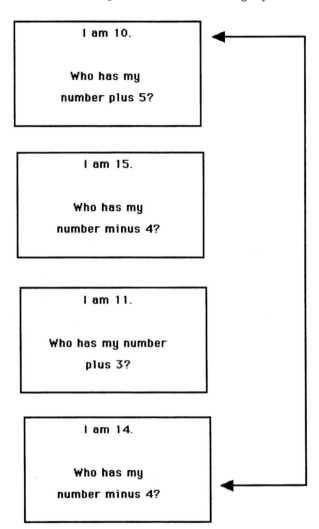

Figure 20. MentComp cards are made by asking a question on the bottom part of a card which is answered on the top part of the next card. The final question is answered on the top part of the first card.

My number is 14. Who has my number plus 4?
My number is 18. Who has my number minus 3?
My number is 15. Who has my number plus 2?
Carol Smith
First Grade

Multiplication and Division MentComp Activity:

Skills: Basic multiplication and division facts

Each of the following lines is printed on an individual card. Comic figures or cartoon characters can be added for visual interest.

My number is 5. Who has my number times 2?

My number is 10. Who has my number divided by 5?

My number is 2. Who has my number times 4?

My number is 8. Who has my number times 2?

My number is 16. Who has my number divided by 4?

My number is 4. Who has my number times 3?

My number is 12. Who has my number divided by 4?

My number is 3. Who has my number times 2?

My number is 6. Who has my number times 3?

My number is 18. Who has my number divided by 2?

My number is 9. Who has my number times 8?

My number is 72. Who has my number divided by 72?

My number is 1. Who has my number times 21?

My number is 21. Who has 8 times 4?

My number is 32. Who has 7 times 5?

My number is 35. Who has 7 times 7?

My number is 49. Who has 9 times 3?

My number is 27. Who has 3 times 5?

My number is 15. Who has 8 times 5?

My number is 40. Who has 7 times 6?

My number is 42. Who has 8 times 8?

My number is 64. Who has 8 times 7?

My number is 56. Who has 6 times 8?

My number is 48. Who has 4 times 9?

My number is 36. Who has 9 times 5?

My number is 45. Who has 9 times 6?

My number is 54. Who has 45 divided by 9?

Linda Abbott

Fourth Grade

Adding, Subtracting, Multiplying and Dividing Integers MentComp Activity:

Skills: Introduction to Algebra

Perform the four basic operations using signed numbers.

Item	Answer	Question
1.	10	$-8 \div 4$
2.	-2	-3×-8
3.	24	$-15 - 7$
4.	-22	$-12 + 17$
5.	5	$-3 + 7$
6.	4	$3 - 10$
7.	-7	$-2 \div -1$
8.	2	$-2 + -6$
9.	-8	$-7 - -3$
10.	-4	12×-3
11.	-36	$-4 + 11$
12.	7	$5 - -11$
13.	16	$-4 - -8$
14.	-12	$-16 \div -2$
15.	8	-7×3
16.	-21	$-26 + 12$
17.	-14	$-18 \div -6$
18.	3	$15 \div -5$
19.	-3	$-9 \div 1$
20.	-9	-8×-4
21.	32	$-16 - -1$
22.	-15	$-23 - 17$
23.	-40	$-18 + 18$
24.	0	$-12 + 22$

TACTUAL, VISUAL, INDIVIDUAL
OR SMALL GROUP ACTIVITIES

Task Cards

Task cards are simply puzzle cards with questions and answers prepared by the teacher. Shapes are cut from posterboard, tagboard or construction paper, a question or problem is written on one portion of the shape and the answer is written on the other portion of the shape. Then each shape is cut into two pieces and no two shapes are cut apart in the same way. The result is a collection of puzzles which the students place together to identify the correct question and answer. The individual puzzles are self-checking since the puzzle pieces fit together (see Figure 21). These materials are appropriate for the tactual learners who need the small muscle movement and the visual learners who need to see the material. Sociological groupings can also be accommodated by having the students work alone, in a pair, or small groups. If these cards are used in small groups, the students can discuss the concepts and skills which will also benefit the auditory learners.

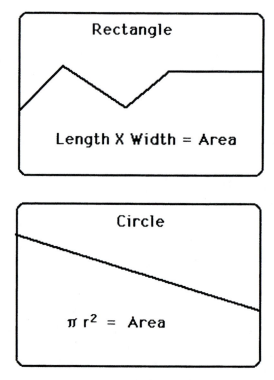

Figure 21. Task cards are made by writing a question or problem on one section of a card and the answer on the other section. Then each card is cut into two pieces with no two cards being cut apart in the same way.

Specific Task Card Activities

Money Task Cards:

Skills: Identifying pennies, nickels, dimes, quarters and their values
Cut poster board into strips or shapes. Cut out pictures of coins from an old workbook, glue coin groups of your choice to one end of the poster board. Write the correct money amount on the other end. Cut each strip into a two piece puzzle. The students will match the appropriate groups and values. Rubber stamps of coins can also be used to make task cards.
Felicia Brown
First Grade

Ordinal Numbers Task Cards:

Skills: Identify ordinal positions first to tenth
Write the ordinal position number words first to tenth on one end of the posterboard. Write the numerals 1–10 on the other end. Pictures depicting the positions could also be used with this activity.
Felicia Brown
First Grade

Geometric Shapes Task Cards:

Skills: Identify the square, circle, triangle, rectangle, trapezoid, rhombus, and parallelogram.
Use index cards or sections of poster board to draw geometric figures on one end and write the figure's name on the other end. Cut each card into a two-piece puzzle.
Laminate the cards for durability.
Have the students match each figure with its name by placing the pieces together.
Kirk Thompson
Special Education

Algebra I Task Cards:

Skills: Equations in One Variable
Solving equations using one or more operations.
Solving equations with variables in both members.
Cut and laminate the separate task card pieces. An overhead writing pen can be used to write on each side of the task card. This way, when one activity is completed, the cards can be re-used. Create your own style of task cards by varying the shape or size.

Sample Problems:

$-73 = 13 - h$ $h = 86$
$\frac{4}{5} t = 8$ $t = 10$
$-144 = 16n$ $n = -9$
$b - 32 = -82$ $b = -50$
$3c = 21$ $c = 7$
$4v - 7v = 21$ $v = -7$

Dena Bradway and Tracy Markham
Eighth and Ninth Grades

Checking Subtraction With Addition Task Cards:
Skills: Use addition as the checking operation of subtraction
The students will match the addition problems that check the appropriate subtraction problems. Write subtraction problems on one end of the card and the addition problems that check the subtraction problems on the other ends of the cards before cutting into puzzle pieces.

$$\begin{array}{r} 42 \\ -\ 29 \\ \hline 13 \end{array} \qquad \begin{array}{r} 29 \\ +\ 13 \\ \hline 42 \end{array}$$

Gloria Clay
Third Grade

Addition Task Cards:
Skills: Add three one-digit numbers
Place three one-digit numbers on one end of the card and the appropriate sum on the other end before cutting the cards into puzzle pieces. The game can also be made more difficult by cutting these into three puzzle pieces with two addends, another addend, and the sum on each of the pieces.
Gloria Clay
Third Grade

TACTUAL, VISUAL ACTIVITIES

Electroboards

Electroboards are made by using two sheets of posterboard of any size or shape. File folders can also be used to provide the front and back of the electroboard. Other necessary materials and equipment needed are: (1) aluminum foil, (2) masking tape, (3) hole punch or leather tool punch, and (4) rubber cement or hot glue.

Step One: Questions and answers are written on the front of one piece of the posterboard. Designs can be also added for esthetic appeal. The questions and answers are written in a random manner creating a matching style format (see Figure 22).

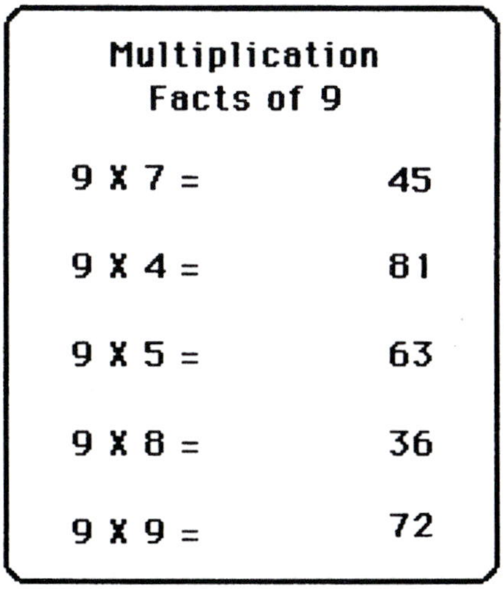

**Multiplication
Facts of 9**

9 X 7 =   45

9 X 4 =   81

9 X 5 =   63

9 X 8 =   36

9 X 9 =   72

Figure 22. Questions or problems and the appropriate answers are written in random order on the front of one piece of posterboard.

Step Two: Both pieces of poster paper are laminated separately.

Step Three: A hole punch is used to make holes beside questions and the answers (see Figure 23).

Step Four: The first piece of posterboard with the questions and answers is turned over. A strip of aluminum foil is folded into eight thicknesses. The length of the strip should be sufficient to match a question hole and the appropriate answer hole. Connect the first pair of corresponding question and answer holes by completely covering the holes with the aluminum foil strip. Use masking tape to cover the foil strip completely. The masking tape acts as an insulator.

Step Five: Connect another corresponding question and answer hole with foil and insulate with the masking tape before connecting another pair of question and answer holes. Insulating each strip of foil with masking tape before using another foil strip is very important. Continue

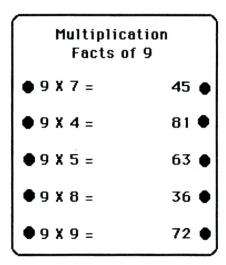

Figure 23. A hole punch is used to make holes beside the questions or problems and the answers.

the foil and masking tape technique until all pairs of holes are connected (see Figure 24).

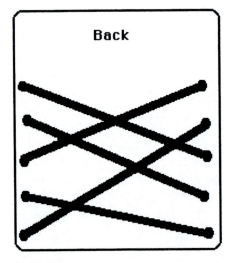

Figure 24. Connect corresponding question and answer holes by completely covering holes with an aluminum foil strip. Then the foil strip is completely covered with masking tape. The process is continued until all pairs of holes are connected.

Step Six: Once the electroboard front is completed, hot glue the two pieces of posterboard together to cover the aluminum foil and masking tape strips.

Step Seven: A continuity tester, purchased from an auto supply store, is used to determine if a circuit is connected. The tester is used with the electroboard to provide immediate feedback to the student's choice of answers. When the two ends of the continuity tester are placed on the matching question and answer holes, a circuit will be made and the light will shine on the tester. Some continuity testers have very pointed ends which should be rounded or dulled before using them in a classroom.

Specific Electroboard Activities

Geometric Shapes Electroboard:
Skills: Identify geometric shapes
Cut out two cylinder shapes from poster board for the electroboard. Cut geometric shapes from old workbooks or from sheets of construction paper. Place shapes on one side of the electroboard and the names of the shapes on the other side. Follow directions for making electroboard. The students will match the appropriate names and shapes.
Vera Lucas and Felicia Brown
First Grade

Addition Electroboard:
Skills: Add a one-digit number to a two-digit number without regrouping.
Subtract one-digit numbers from two digit numbers without regrouping.
Select problems from a text or create your own. Follow directions for making electroboard. The students will match the appropriate answers and problems.
Felicia Brown
First Grade

This can be done for any level of addition, subtraction, multiplication, or division problem. As the problems become more difficult, students may use paper and pencil to complete each problem and then use the electroboard to check their problems.

KINESTHETIC, VISUAL, MOBILITY, INDIVIDUAL OR SMALL GROUP ACTIVITIES

Large Floor Games

Large floor games are especially appropriate for kinesthetic students and students who have a high need for mobility. Large floor games are easily made from inexpensive shower curtains, permanent markers, construction paper pieces cut from specific patterns, and a hot glue gun. These activities can also be produced on posterboard for use with tactual students who need more small muscle movement.

Specific Large Floor Games

Hundred Chart Floor Game:
Skills: Identify numerals 0–100
Orally read numbers to 100
Rote counting to 100
Identify the numeral that comes before, after, and between given numerals to 99
Count and write by 2's and 5's to 100
Draw 100 footprints on a shower curtain or cut footprints from construction paper, laminate them and glue them to the shower curtain with a hot glue gun. The numbers should be arranged in groups of ten as in any Hundred Chart. Each row of ten numbers could be in a different color to assist the global learner in remembering the various numbers. Number the footprints from 1–100. The students can walk on the footprints starting at 1 and going to 100. The teacher or other students can also cover certain numbers with construction paper to identify what comes before and after, the 2's, 5's, odd numbers, even numbers, or other patterns of numbers on the chart.
Carol Goldsmith
First Grade

Coin Values Floor Game:
Skills: Identify the penny and its value.
Identify the nickel and its value.
Identify the dime and its value.
Identify the quarter and its value.
Divide a shower curtain into squares. Glue a large picture of a coin in

each square. The students will place the "game board" on the floor, toss a bean bag, and identify the coin the bag lands on and name its value.

Carol Goldsmith
First Grade

Fraction Match Floor Game:
Skills: Match fractions with denominators of four or less with appropriate shaded regions or portions of sets

Divide the shower curtain into sections. Draw various geometric shapes divided into fractional parts into each section. Shade various fractional parts of each shape (ex: ¼, ⅓, ½, ⅔, ²⁄4, ¾, etc.). Make answer cards with the appropriate fractions on them. Students toss a bean bag on the gameboard and then identify the appropriate fraction.

Gloria Clay
Third Grade

SMALL GROUP ACTIVITIES

Circles of Learning

Circles of learning are used with small groups of students, approximately 5, and questions for which multiple answers are possible. The students are divided into groups and a recorder is assigned. (See Chapter 3 for other roles.) The students are presented the question and respond one-at-a-time going clockwise around the circle of students. The students respond at their identified times, but may pass if they do not have an answer for any turn. The purpose is to identify as many correct answers as possible during the identified time (usually 2 minutes). The recorder also responds as s/he writes all of the responses given on a sheet of paper. After the time limits have expired, the answers provided by the students are compiled by the teacher using the overhead projector or chalkboard. Each group responds in turn, providing one answer at a time. If the answer has already been presented by another group or if the answer is incorrect, the group loses a point. All correct answers receive one point and the group which accumulates the most points wins (Dunn, 1988).

Specific Circles of Learning

Questions appropriate for Mathematical Circles of Learning include, but are not limited to:
1. What two numbers added together equal 16?
   What two numbers added together equal 10? etc.
   Explain to the children that you are looking for every possible combination.
2. Name some objects around the room that are shaped like a circle, square, etc.
3. Name even numbers that have the digit 2 in the ones place.
4. Identify geometric shapes and the area formula for each.
5. Identify fractions equivalent to one-half.

Team Learnings

Team Learnings are another small group activity for approximately 5 or 6 students. The students are divided into groups and roles are assigned (see Chapter 3). They are given a selection of material to read based on any mathematical concept. Put the material you want the students to learn into simple, clear sentences; the material should be no more than a few paragraphs. Then add three different types of questions related to that content. The first type of questions should be factual and based directly on the material, so that the youngsters can easily find the answers and feel successful. The second type of questions should be answerable only through thinking—analysis, comparison, extrapolation, and so forth. Their answers should not be in the material. The third question—or objective—should require that the team take the new information taught through the material and use it in a creative way by making something, such as a poem, a pictorial representation, a composition, a three-dimensional puzzle, a scramble board, a set of task cards, a word-quiz, or an original song or rap. Once the Team Learning is completed, the answers are shared with the entire class (Dunn, 1988).

Specific Team Learning

Team Learning on Quadrilaterals:

Team Members:

1. _____   4. _____
2. _____   5. _____
3. _____   6. _____

Recorder: _____

Read the following passage and then reach consensus with the other group members and answer all questions. Go with the majority in terms of disagreement.

Quadrilaterals are polygons with four sides. These sides are line segments which intersect at the endpoints of each segment. Each one of the intersections is known as a vertex.

Parallelograms are one kind of quadrilateral. These are distinguished by the fact that both pairs of opposite sides are parallel. One special parallelogram is the rhombus because its sides are all congruent.

Another type of quadrilateral is a rectangle. Rectangles are set apart due to the fact that they have four ninety degree angles. A special rectangle with four congruent sides is a square.

The last type of quadrilateral to be discussed is the trapezoid. A trapezoid must have one pair of parallel sides. The angles and the other two sides of a trapezoid do not have to meet any special requirements.

Questions:

1. How many sides do all quadrilaterals have?
2. Describe parallelograms, rectangles, and trapezoids.
3. List any quadrilaterals that could be considered parallelograms.
4. Why are all the quadrilaterals mentioned not considered to be parallelograms?
5. Create a rhyming poem or rap, or build a mobile or a mini bulletin board to explain quadrilaterals.

VISUAL, AUDITORY, MOBILITY, WHOLE GROUP ACTIVITIES

Review Activities

Review Trivia Basketball:

Skills: Any skills or concepts identified by the teacher

The student will correctly answer a mathematics question related to a specific mathematics concept or skill, and, in order to acquire points for that question (and his team), must make a basket by throwing a nurf basketball through the hoop or by throwing a heavy wad of paper into the trash can at the front of the room. Divide the class into two teams, and pair students to compete against each other. Show a question on a section of posterboard or an overhead transparency and ask the question directed at those two students. The first one to slap his/her desk gets a chance to answer the question. If the answer is correct, s/he may try to shoot a basket for 2 points. If s/he misses the question, the other student gets a chance to answer. If s/he is also incorrect, explain the correct answer and give those same students another question. When the question is answered correctly, the student tries to obtain two points for his team and then two additional students are selected. Provide a positive incentive for the winning team, such as a small treat or new pencils, etc. Sometimes this can be used before a holiday weekend and trivia questions from all subjects can be asked.

Tracy Markham
Ninth Grade

Jeopardy Review:

Skills: Review of any mathematical content

Sheets of posterboard are divided into halves or fourths and answers to mathematical questions or problems are printed on the cards. The individual cards are held up and the first student to raise his/her hand has a chance to provide a correct question to the answer. As in the television game, all responses must be in the form of a question or the response is not correct. The student who responds correctly, in the form of a question, is given a simple reward. The rewards can be stickers, book marks, or any token identified by the teacher. Teachers should realize that there will often be more than one correct question for the identified answers and should accept any correct question presented with sound reasoning. If the student response is questionable, the student should be asked to explain his/her reasoning.

CONCLUSION

These activities are presented as only a sampling to provide examples for teachers to use. Teachers are encouraged to adapt the activities and create their own to accommodate the learning style needs of students when short periods of drill and practice are required. Drill and practice sessions are needed for students to master various mathematical concepts and skills. However, this practice should be for short periods of time and in varied formats. Endless worksheets are boring to the students and do not meet the learning style needs of all the students. Furthermore, if lengthy drill and practice sessions are used, students may repeat error problems and have difficulty learning the correct way to work the problems.

Chapter Seven

PORTFOLIO ASSESSMENT IN MATHEMATICS

NEED FOR CHANGE IN ASSESSMENT

L earning style teaching advocates the use of a variety of teaching methods and strategies that are nontraditional. In many instances, the variety of visual, auditory, tactual, and kinesthetic activities utilized through learning style based teaching make assessment through the traditional paper and pencil tests difficult. These alternative teaching methods require means of alternative assessment. States such as Kentucky, Connecticut, and Rhode Island have been examining alternative means of assessing learning, and the use of portfolios has emerged as a leading strategy. In fact, Vermont and Rhode Island now utilize portfolios for all state assessments, and plans are underway for California and Kentucky to follow (Brandt, 1992; Maeroff, 1991).

Furthermore, the *NCTM Curriculum and Evaluation Standards for School Mathematics* (1989) addresses the need for authentic assessment which is quite different from the way most teachers currently assess student learning. "The main purpose of evaluation [should be] ... to help teachers better understand what students know and make meaningful instructional decisions" (NCTM, 1989, p. 189). Assessment strategies should focus on instruction and student learning, what the students know as opposed to what the students do not know or understand. This assessment includes informal observations by the teachers, which is a constant part of effective mathematics instruction. Furthermore, assessment is no longer seen as being separate from instruction, but is now an integral part of the learning process which involves students in establishing goals, self-assessment, and choosing appropriate evaluation techniques (Stenmark, 1991).

Since the way mathematics is taught must change to incorporate learning styles and to prepare students for a more technological world in which decision making and problem solving are emphasized, assessment of students must also change. Any assessment activities used in the classroom should closely resemble learning tasks, be meaningful to the students and have instructional value for the students when prompt feedback is received (Rindfuss, 1993). Furthermore, the use of calculators, computers, and manipulatives must be included in assessment to correspond to effective instruction in mathematics. Instead of focusing on isolated mastery of skills, a more holistic view of mathematics which incorporates a broad range of topics is deemed necessary. Teachers should utilize multiple forms of assessment including demonstrations, investigations, enhanced multiple-choice items, and open-ended questions, as well as oral and written activities to get a picture of the whole child (Fox, 1993). With this variety in assessment formats, a variety of scoring techniques is also required. Rubrics which facilitate scoring of students' work, record keeping on note cards or on a computer, and maintaining a portfolio of student work can facilitate understanding of what the students know. The emphasis must shift from reliance on a single form of assessment, typically a standardized test, to the use of "multiple assessment techniques that are aligned with the curriculum" (NCTM, 1989, p. 192). One way to organize these multiple assessments is through the use of portfolios.

AN OVERVIEW OF PORTFOLIO ASSESSMENT

Teachers are often perplexed or confused by the term, "portfolio assessment," but this somewhat new terminology can be more readily understood when work portfolios professionals often assemble to display their talents and capabilities are considered. Student portfolios are similar in nature; they are a way to showcase student work including various types of assignments, writings, reports, and projects (Stenmark, 1991). However, a portfolio is more than a collection of samples of a student's work. The portfolio is organized to demonstrate growth in mathematical performance. To be most effective, it is a means of longitudinal assessment of the internalized mastery of knowledge (Farr, 1993), "a summative evaluation of student success and growth" (Fox, 1993). Authorities such as Howard Gardner recognize the different intelligences and the styles in which new learning is acquired. The various kinds of intelligence and learning styles of students are often not considered in the delivery of the school curriculum or the assessment means (Herbert, 1992). The portfolio helps "track and evaluate students' performance over time, pinpoint their strengths and weaknesses, examine their learning strategies, and assess their attitudes" (Tierney, 1991, p. 63) in the confines of the different kinds of intelligences and learning styles. "A portfolio, then, is a portfolio when it provides a complex and comprehensive view of student performance in context" (Paulson, Paulson, & Meyer, 1992, p. 63).

The concept of portfolio assessment might be unfamiliar to the students as well as the teachers. Therefore, training in the specific methods and techniques is advantageous to teachers beginning to use the portfolio method of assessment. When portfolio assessment is introduced to the students, a brief introduction by the teacher could include samples of his/her personal portfolio or samples of loaned portfolios from former students. The use of portfolios by professional artists could also be explained. The activities described are more attuned to the global method of instruction. The analytic student might benefit by a list of items to be included and a rating scale scoring system (rubric) if the school district utilizes a letter grade for the reporting system. The analytic student, many times, functions best when explanations of the rules and regulations are stated at the first of a project instead of being allowed to evolve during the course of the project. In fact, "the student should be provided with scoring criteria and models of excellent performance or production as part of instruction" (Wiggins, 1992, p. 29). After examples of scoring criteria and grading or rating scales used by other teachers are examined, then the students could assist in developing the rating scale and/or scoring criteria in an attempt to build ownership of the evaluation. In school districts where the portfolio is used in lieu of major tests, decisions must be made regarding grade assignment and required level of mastery. The rubrics and scoring criteria would be used to determine if the student had demonstrated mastery of the content. These scales can also evaluate life-skills such as time management, work skills, and social group interaction skills. While portfolio assessment is a holistic approach usually desired by global learners, the entire class of analytic and global learners should receive a positive introduction to the concept. "Ideally, the steps leading to an assessment and the assessment itself will be a learning experience for students" (Maeroff, 1991, p. 280).

A centralized station for use in portfolio assembly, as well as regularly scheduled time for portfolio work and teacher conferences are required for the successful use of portfolio assessment. The portfolio can be housed in a folder, pocket folder, box, notebook, or expandable file. There should be one area of the center devoted to the collection of the samples while another work area would house the actual portfolios. All work, checklists, and observation forms should be dated to aid in assessing growth.

The students should feel ownership of the portfolio. Therefore, the containers should be decorated by the students. Pictures may be drawn or pasted onto the surface, and enhancement items such as novelty paint, lace, trim, sequins, and other art media could be applied. Time should be allowed during the class period for the creative endeavor. The teacher's maintaining a personal portfolio adds to the students' interest. Included in the teacher's portfolio would be items similar to the students' work: (1) a log of personal experiences in real-life mathematical situations, (2) a response page to reflect feelings and thoughts, (3) goals for growth for the period, (4) lists of projects to complete or books to read for professional growth, (5) videotapes or audiotapes of completed projects, and (6) an introductory letter to explain the selection of portfolio contents.

The contents of the portfolio should include examples of authentic assessment projects in which the student has utilized a combination of knowledge and critical/creative thinking skills to solve problems. Since authentic assessment has become an important concept for performance-based evaluation, completing real-life activities involving mathematics and problem solving skills is important. Emphasis should be on solving meaningful, real-life problems instead of disconnected factual problems which demonstrate mastery of basic computational skills. By utilizing a variety of assignments, the students' interests are accommodated and provide opportunities for assessing the learning styles of the student which increase motivation, persistence, and responsibility (Berger, 1991; Chittenden, 1991).

Accessibility to the contents of the portfolio should be restricted to the student, teacher, and parent unless permission is granted by the student him/herself or ground rules for sharing have been developed. "Students need to know who is going to see this work so they can make decisions . . . about whether to include personal feelings" (Chow, 1992, p. 4). The sharing of the contents with other students can be a profitable, educational experience. If the class does utilize a sharing time, students should develop guidelines and techniques for positive, constructive evaluation. A sheet for Portfolio Comments can be included at the back of the portfolio for viewers to respond to identified items (Hansen, 1992).

PORTFOLIO CONTENTS

The focus of student portfolios should be on "student thinking, growth over time, mathematical connections, student views of themselves as mathematicians, and the problem-solving process" (Stenmark, 1991, p. 37). Therefore, a wide variety of activities are deemed necessary for inclusion in the collections.

Samples of Activities Included in Portfolios:
1. Three or four objective goals
2. Examples of authentic assessment projects
3. Performance based assessments
4. Enhanced multiple choice items
5. Open-ended questions
6. Word problems created by the student
7. A photography or sketch of student work
8. Videotapes
9. Audiotapes
10. Student explanations
11. Error analysis
12. Reports of group work
13. Work from another subject area
14. Art projects
15. Teacher observation checklists
16. Student selected samples of work
17. Teacher selected samples of work
18. Samples of journal entries
19. Newspaper and magazine articles
20. Participation charts and frequency counts
21. Learning styles checklists
22. Student interest inventories
23. Mathematical autobiography
24. Cover letter
25. Recognition of student achievement
26. Teacher response pages and checklists

Objectives

Three or four objective goals are selected during a conference between the teacher and the student. At each conference, the student and the teacher should select a few objectives to be the goal of achievement for the grading period. A standardized list of objectives appropriate for the grade level based on a state curriculum guide or the *NCTM Standards* (1989) is advantageous to use. In fact, this type of standardized list of objectives is required in states beginning to use portfolios in lieu of standardized state mandated testing.

Authentic Assessment Projects

Examples of authentic assessment projects in which the student has utilized a combination of knowledge and critical or creative thinking skills to solve real life problems are included in the portfolio to demonstrate mastery of the process of mathematics as opposed to just speed and accuracy. These projects should be interesting to the students to promote persistence, be thought provoking, develop a variety of thinking styles, involve processes as opposed to skill mastery, and be open. A variety of tasks and activities appropriate for this type of assessment are provided in the *NCTM Standards* (1989) or teachers could create their own assessment projects. These include performance-based tasks, investigations, and "draft, revised, and final versions of student work on a complex mathematical problem, including writings, diagrams, graphs, charts, or whatever is most appropriate" (Stenmark, 1991, p. 37).

Performance-Based Assessment

Performance-based assessments are used for student demonstration of knowledge of a mathematical concept by completing a specific task. Modeling a mathematical concept with manipulatives or developing and recording steps in problem solving are excellent performance based tasks to use (Fox, 1993). The process the students use to complete the task is as important as the completed task or product. Documentation of these activities may include videotapes, audiotapes, or written descriptions by the teacher or other students. Students who may have difficulty with paper and pencil tasks can often demonstrate knowledge through a hands-on activity. Furthermore, as manipulatives are used in mathematics instruction, they must also be included in assessment. One example of such a task is to give the students base ten blocks and ask them to demonstrate how to solve a particular problem. The teacher could then use a scoring scale or rubric to assess student performance (Fox, 1993).

Enhanced Multiple Choice

Enhanced multiple choice items are similar in nature to the traditional multiple choice items found on many standardized tests. However, the enhanced items provide for an explanation of the students' thinking. For example (Rindfuss, 1993):

Enhanced Multiple Choice Example:
Using quarters, nickels, and pennies, what is the least number of coins needed to make 64 cents?

A. 9
B. 8
C. 7
D. 6

Explain why your answer is correct.

Open-Ended Questions

Open-ended questions have more than one correct answer and can be solved in a variety of ways. For example (Rindfuss, 1993):

Open-Ended Question Example:
Place the following numbers in two sets and describe the rules you followed for their placement, 3, 4, 6, 9, 11, 15, 16. A Venn Diagram could be used for the placement of the numbers. Then the students could be asked to arrange the numbers in two other sets and again describe the rules used for the placements.

Word Problems Composed By Students

Word problems created by the students incorporate higher level thinking skills and demonstrate mastery of specific problem solving techniques and strategies. This creative writing based on real-life experiences can also address specific criteria provided by the teacher.

Example of Directions for Creating Word Problem:
Create a word problem following these guidelines:

1. The problem must be a two-step problem.
2. The problem must involve percentages in some way.
3. The problem must depict a "real world" situation.

Photographs or Sketches of Student Work

A photograph or sketch of student work documents the student's work with manipulatives, calculators, computers, or with models of multi-dimensional figures (Stenmark, 1991). This incorporates the use of instructional activities with assessment and can be used in combination with student explanations of the processes completed.

Videotapes

Videotapes of performance based tasks such as working with manipulatives could also be added to the portfolios to give a more complete picture of achievement in mathematics. Videotaping final reports or projects of cooperative learning groups could also increase the clarity of the results of the group.

Audiotapes

Audiotapes of student reports, raps, rhymes, jingles, poems, and the like provide an opportunity for the students to use mathematical content in a new and different way. Once the content has been covered in class, the students can create stories, commercials, raps, jingles, or poems using the information. If the students are allowed to use the information creatively, long term retention may be increased as student attitudes regarding mathematics are improved. Videotapes of students performing their creations could also be included.

Student Explanations

Student explanations involve writing skills to present a substantiation of how a particular type of problem is worked. Students can also "defend in writing how they answered a [specific] math problem" (O'Neill, 1992, p. 15) or write a letter to a classmate that was absent so that s/he will understand what was done in class that day. When students explain a process in this way, the teacher can easily determine if the process has been mastered by the student or if additional instruction is needed.

Error Analysis

An error analysis completed by the student explains why any identified problems were worked incorrectly. The student may be able to identify that a careless mistake in a basic operation was made or that a particular step in the process was omitted. However, the error analysis may also reveal that the student does not know what is wrong with the problem. The papers of this type which are included in portfolios typically demonstrate the student's correction of errors or understanding of previous misconceptions (Stenmark, 1991).

Reports of Group Work

Reports of group work completed by the student should include the individual's contribution to the overall project (Stenmark, 1991) and reactions to the group process.

Work From Other Subject Areas

Work from another subject area could also be included in the portfolio to relate mathematics to other curriculum areas and foster mathematical connections as advocated in the *NCTM Standards* (1989).

Examples of Curriculum Correlation:
Activities of this type might include increasing or reducing a recipe in home economics, drawing a house plan in vocational education, collecting data about a science or social studies topic and presenting the information in graph form, comparing currency exchange rates, writing book reports which connect literature and mathematics, or comparing gravity on the various planets in graph form, as well as computer demonstrations.

Art Projects

Art projects can foster spatial perception skills and document understanding of various mathematical concepts (see Chapter 5).

Examples of Art Projects:
Projects which foster spatial perception skills might include scale drawings of maps, Origami projects, coordinate pictures, triad rota-

tion pictures, string or line art, tessalation drawings, or kaleidoscope pictures.

Coordinate pictures involve the use of coordinate numbers plotted on a graph to form a picture, while triad rotation pictures are three-dimensional cubes, boxes, and cylinders rotated and colored to form a pattern. String art or line design results in geometrical shapes which are formed through connecting points with string or with straight lines. Patterns evolved through coloring various geometrical shapes are tessalations drawings, and kaleidoscope pictures are produced on one segment of a circle and reproduced on the other equal parts of the circle.

Teacher Observation Checklists

Teacher observation checklists which focus attention on the student's work habits are also included in the portfolios (Forester & Reinhard, 1989). The student's work skills, time management skills, and social interaction skills would be assessed as they related to the completed assignments. The checklists make identifying and assessing specific information about students more efficient when finding the time to write anecdotal notes about all of the students in a particular class is difficult. Each day a few students could be observed; the focus should be on the entire picture of student achievement and, therefore, multiple observations of the same students are required. Students should be observed when working independently, in small groups, and in whole group instruction since students may be more productive in one sociological grouping than the others employed by the teacher. Videotapes of classes could be used to facilitate the observations and completions of checklists (Stenmark, 1991). The students might also be given a copy of the checklist to complete to assist in isolating problem areas or in identifying strengths.

Student Selected Samples

Student selected samples of work should be included to demonstrate learning and understanding deemed important by the students. Part of the underlying philosophy behind portfolio assessment includes involvement of the students. No longer is assessment completed solely by the teacher. The teachers and students identify goals, strengths, and areas where improvement is needed. Therefore, the student becomes an integral part of the assessment process (Stenmark, 1991). As students select samples of work for inclusion in the portfolio, a personal statement by the student should be attached to each sample. This should include the student's reasoning for selecting the work (Knight, 1992) and can provide valuable insight for the teacher.

Teacher Selected Samples

Teacher selected samples of work are also included in the portfolio and clearly identified by a T at the top of each selection (Stenmark, 1991). These additional entries provide samples of the student's work which the teacher values as demonstrating growth or a need for growth by the student.

Samples of Journal Entries

Samples of journal entries or logs completed by the student should demonstrate reflection about mathematics by the student. Including the entire journal in the portfolio is not necessary, but samples of the writings should be included to provide a clearer picture of student understanding (Fox, 1993).

Examples of Journal Entries:
These entries could incorporate student reactions to mathematics used in the world outside of the classroom, explanations of processes used in solving a problem, personal definitions, expressions of feelings concerning student learning, or answers to specific questions identified by the teacher.

Newspaper and Magazine Articles

Newspaper and magazine articles which include mathematical information could be added by the student to increase the awareness of how mathematics is used by people on a daily basis. Students' reactions to each of these entries would allow reflection and reaction to mathematics used in a variety of situations.

Participation Charts and Frequency Counts

Participation charts and frequency counts provide another avenue for documenting student activities. Classroom experiences which involve manipulatives, role play, or cooperative learning can be listed with the date completed and the student's reaction to the task. Rating scales or rubrics assessing the specific work completed could also be included.

Learning Styles Checklists

Learning styles checklists provide readers of the portfolio, especially parents, an understanding of the learning style profile of the student. If an assessment of learning style preferences has been administered, the result of the inventory or an informal checklist of learning style characteristics could be placed in the portfolio.

Student Interest Inventories

Student interest inventories may also be included in the portfolio since student interests play an important role in the selection of the activities for assessment. These inventories also provide valuable information for teachers and parents who wish to more actively engage students in the learning of mathematics.

Mathematical Autobiography

The mathematical autobiography addresses the many experiences remembered by the student which involve mathematics. This autobiography focuses on the student's associations with mathematical concepts and his/her reactions to specific activities and prior experiences. The information provided in the autobiography provides valuable insight to student attitudes regarding mathematics.

Cover Letter

A cover letter or summary of the portfolio is written by the student after the collection is complete and placed at the front of the portfolio for reviewers to read. This cover letter may include explanations of the contents, any growth the student has observed in him/herself, questions or problems the student still has with any particular concept, and identification of projects or contents which the student feels are incomplete or need improvement (Stenmark, 1991). This letter provides an overview of the entire portfolio process as perceived by the student.

Recognition of Student Achievement

Recognition of achieved goals or demonstrated strengths should be included to commend the completed work. A certificate of achievement or a note written by the teacher would be one way to meet this need. A response sheet could also be used for comments to be written by students, parents, or others who view the portfolio.

Teacher Response Pages

Teacher response pages and checklists are provided to address the quality of work included in the portfolio as well as suggestions for improving the portfolio and individual student assignments.

ACCOMMODATING LEARNING STYLES THROUGH PORTFOLIOS

A variety of tasks placed in a portfolio incorporate visual, auditory, tactual, and kinesthetic skills. The various projects may be basically of one learning style preference, but may also include a combination of several learning style elements. These activities include:

1. Journal entries or logs—visual and tactual,
2. Photographs of manipulative work—tactual, kinesthetic, and visual,
3. Graphs constructed from surveys—visual, auditory, tactual or kinesthetic based on the activity,
4. Cassette tapes of raps, jingles, or poems which incorporate mathematical content or interviews—auditory,
5. Videotapes of mathematical simulations and projects—visual, auditory, tactual, or kinesthetic based on the completed activity,
6. Photographs of bulletin boards created by students—visual, tactual, and kinesthetic,
7. Art projects fostering spatial perception skills—tactual and visual,

8. Interdisciplinary projects—visual, auditory, tactual, or kinesthetic depending on the selected activity.

PORTFOLIO ORGANIZATION

The portfolio needs to have some type of organizational pattern. Students' learning styles are important to consider in relation to organization. Analytic students tend to categorize materials in an orderly manner, while global students "chunk" file. A simple method to assist students could aid in maintaining a portfolio that clearly displays its contents.

Simple dividers can be made by folding large sheets of construction paper in half to make file folders. Using different colors of paper can assist the student in locating and filing the samples. One section could contain checklists, inventories, personal assessment, and anecdotal records. A second section could contain formal assessment means such as results or samples of specific tests. A final segment could be reserved for samples of work. The sample section would be revised and purged periodically, while the other two segments would remain more constant. Students would have the option to prepare more organized portfolios if they desire.

The act of purging the files gives the student the opportunity to analyze his/her work and make decisions concerning the inclusions of samples of varying quality. The teacher needs to have provided guidance and suggestions in the criteria for evaluation of the portfolio and procedures to be used. Care should be given to emphasize the importance of including samples which reflect student growth over a period of time.

Since the purpose of portfolio assessment is to evaluate student growth over a specified period of time, similar assignments, questions, and projects should be included periodically over the designated time period. Regularly adding checklists, interviews, and the other activities which are dated can add to the organization of the portfolio as well as assist in identifying student achievement.

EVALUATION OF PORTFOLIOS

The evaluation of the portfolio should include analyzing both process and product and is usually formative in nature. Process growth is measured through checklists, observations, and anecdotal records. Checklists of specific objectives also aid the teacher in focusing on what the students do rather than the lesson plans implemented (Forester & Reinhard, 1989). Samples may be utilized in a formal assessment of the product; however, multiple samples (six to eight) of the same tasks need to be assessed in order to provide validity. As the portfolios are analyzed in a formative manner in setting goals for improvement or advancement, they are utilized as more thorough evaluation measures and closely resemble Madeline Hunter's (1989) "glow and grow" teacher/evaluator conferences where outstanding observations are praised and areas of growth are recommended.

In situations where grades are still a school policy, point systems or rating scales are sometimes used to assist in the subjective grade. Knight (1992) "devised a grading matrix and weighted the portfolio grade to be equivalent to about one-fifth of a test grade" (p. 72). Scoring rubrics are also utilized as a means of assigning scores. The rubric contains a scale of possible points and specific criteria for points earned along with examples of work samples assigned the value of locations along the scale (Chow, 1992).

Specific categories for portfolio entries and criteria for evaluating the criteria should be clearly established. These categories and criteria should be ones that are important to the teacher and include qualities that clearly indicate successful work. Only a few criteria which are discussed with the students and included on the front cover of the portfolio should be evaluated at one time. Any identified criteria should be taken into consideration as individual assignments are graded by the teacher (Stenmark, 1991).

The evaluation of the portfolio should focus on three major issues: "habits of mind, projects, and content" (Marzano, 1992, p. 167). On the basis of the portfolio, the teacher assigns grades or writes narrative reports (Wolf, LeMahiew, & Eresh, 1992). After the teacher has previewed the portfolio, a conference with the student should be conducted. The conference aspect of portfolio assessment is both formal and informal. The purpose of the conference is to allow for pupil/teacher interaction in evaluating the work and setting goals for the next grading period. The teacher should structure the conference to allow the pupil to assist in identifying areas needing improvement. The ultimate goal is for the pupil to assume the major responsibility for the assessment and utilize the teacher as a consultant in securing references and/or resources needed to accomplish the newly defined goals. Obviously, this situation is Utopia and may never be achieved even though the master teacher creates the environment to foster the independence. "Self-assessment gives student control. Recording their evaluation . . . transfers ownership to the rightful owner of learning. . . . the student" (Heitterscheidt, Pott, Russell, & Tchang, 1992, p. 73).

PARENTAL INVOLVEMENT

Alternative assessment methods such as portfolios might be new to families; therefore, the teacher should have a plan for introducing the concept to the family. A simple note sent home could serve as an introduction. Portfolios could then be utilized during parent conferences to demonstrate the performance growth of the student. The student might attach a personal letter to his/her family explaining the progress made, frustrations experienced, or the goals set in relation to an attached individual assessment sheet. The teacher would emphasize the positive growth of the students and discuss the goals and objectives jointly developed during student/teacher conference. Parents should have an opportunity to give further input into the plan developed.

One innovative means of sharing portfolios with parents was designed at Crow Island Elementary School, Winnetka, IL (Hebert, 1992).

Sharing Portfolios With Parents:
Portfolio Evenings provide an opportunity for the students to share their portfolios with their parents. Teachers should lead students in advance preparation of the agenda for the events, but the actual portfolio viewing should be directed by the student. Besides providing parents with a longitudinal display of growth, the event offers opportunities for leadership, independence, and self-evaluation for the students.

ADVANTAGES OF PORTFOLIO ASSESSMENT

While the amount of time and effort required to make portfolio assessment a successful experience for the students and the teacher may be seen as disadvantages to its use, the advantages of using portfolio assessment greatly outweigh any limitations of the experience.

Advantages of portfolio assessment:
1. The assessment of the work is made by the student in conference with the teacher.
2. Learning based on portfolio assessment is student-directed.
3. Learning styles based instruction is easily evaluated through a portfolio.
4. Variety is added to the work assignments such as long-term situational problems.
5. The affective domain allows student development of ownership concerning his/her progress.

Discussion of Advantages of Portfolio Assessment

The assessment of the work is made by the student in conference with the teacher. The two evaluate the completed work and make decisions concerning the level of mastery. The initial conference serves as a base point from which the student and teacher set goals. These goals are individualized for the students and allow the teacher to more easily meet the needs of each person in the class. The student and the teacher jointly decide at each conference if the goals set have been mastered, and if so, new goals are established. New techniques or study skills methods might be utilized. Throughout the process of portfolio assessment, the responsibility for the success of the students in meeting the established goals is shared by the student and the teacher.

Learning based on portfolio assessment is student-directed. The students participate in goal selection, evaluation, and providing documentation of their progress. The teachers train the students to "provide evidence of their own learning" (Maeroff, 1991, p. 274). Therefore, the empowerment of the students and teachers provides for a higher level of instruction and assessment than traditional methods have previously allowed.

Learning styles based instruction is easily evaluated through a portfolio. The project work often associated with learning styles based instruction requires a more appropriate means of evaluation than traditional testing of students. The portfolio is an excellent means for validating the different learning approaches (Chow, 1992).

Variety is added to the work assignments such as long-term situational problems (Knight, 1992). The mathematical computation problems are no longer the only source of assessment. The portfolio allows students the opportunity to create their own problems or solve real-life situational problems, which provide for assessment of higher level skills (Szetela & Nichol, 1992).

The affective domain of the portfolio allows student development of ownership concerning his/her progress. Through the cooperative efforts of the teacher and the student, pride and high self-esteem are encouraged (Frazier & Paulson, 1992).

CONCLUSION

As instruction changes to implement the *NCTM Standards* and to incorporate use of learning styles in mathematics, assessment of student learning must also change. Assessment must become an integral part of instruction which examines what students know, understand, and can do as opposed to what students have not learned or can not do. A variety of assessment formats will be required to meet the goals of a holistic approach to mathematics assessment. One way to organize these formats is through the use of portfolio assessment.

While portfolio assessment can be a powerful tool, teachers are urged to ease into this alternative form of assessment. Teachers wishing to implement portfolio assessment should attempt to involve other teachers, parents, and administrators in the process. Gaining ideas from other educators can add in the implementation of this type of assessment. Furthermore, "as more people become aware of the possibilities offered by assessment, greater progress [in portfolio assessment] will be possible" (Stenmark, 1991, p. 60). In addition to the *NCTM Standards* (1989), *Mathematics Assessment: Myths, Models, Good Questions, and Practical Suggestions* by Jean Kerr Stenmark (1991) is an excellent resource for additional ideas, checklists, rubrics, and authentic tasks.

Chapter Eight

CONCLUDING REMARKS

Like master teachers who present introductory, instructional, and then concluding activities for teaching skills and concepts, a few concluding remarks are appropriate to complete what has been presented thus far. These concluding remarks will not summarize all of this text's content, per se, but rather will highlight some of the major emphases.

The use of learning styles to improve mathematics instruction and, therefore, increase student achievement in mathematics was discussed. A learning styles model and suggestions for classroom implementation were given. Teachers were cautioned to start with only one or two elements of learning style as opposed to attempting implementation of an entire model at one time. However, individualizing instruction as much as possible to meet the needs of a diverse group of students was encouraged. In order for students to take charge of their own learning and understand how they can best learn new mathematical concepts and skills, they must be allowed to work through their individual learning styles. The sharing of successful experiences with other teachers was advocated as one way to ease the process of implementing a learning styles based approach to mathematics instruction.

The use of manipulatives to increase mathematical understanding of students was discussed in Chapter 4. This type of instruction enables students to build a good foundation in mathematics which is required before higher level concepts and skills can be learned. Steps to follow in using manipulatives were discussed including modeling the appropriate behavior, discussing and explaining the concept or skill, and recording the abstract version for the students. Teachers were also encouraged to move students gradually from the concrete to the representational stage and, finally, to the abstract stage of mathematical operations. Ways to accommodate learning style preferences while using manipulatives were also presented.

Spatial reasoning as one way to address and reduce gender differences in mathematics was presented in Chapter 5. However, use of proportional and spatial activities were recommended to develop students' abilities to reason mathematically regardless of the gender or the learning style preferences of the students. A sampling of visual activities were also described as categories of visual thinking were presented.

A sampling of activities which can be used to aid retention of mathematical concepts and skills while meeting the learning style needs of a diverse group of students was presented in Chapter 6. Teachers were encouraged to use these activities as models for other activities which cover a wide range of mathematical content, grade level, and learning style preferences.

Portfolio assessment was addressed in Chapter 7. The need for assessment techniques which closely resemble learning tasks and examine what the students know, understand, and can accomplish (as opposed to testing to determine what the students can not do) was addressed. The use of portfolios to organize a variety of assessment techniques and to implement authentic assessment was discussed. Types of activities and assignments, as well as other assessment items, which could be included in portfolios were identified for the reader. Evaluation of portfolios, accommodation of student learning styles, and parent involvement were also addressed.

Due to the limited length of the book, other topics which are seen as critical to effective mathematics instruction were not included. Therefore, readers are encouraged to review additional works to address topics such as cooperative learning in mathematics, problem solving, writing to learn mathematics, and the use of children's literature to teach mathematics. The *NCTM Standards* (1989) should be reviewed and utilized by every teacher of mathematics. However, the use of learning styles when addressing any mathematical content was the major purpose of this writing.

Successful instruction involves a myriad of teacher and student behaviors. Student progress needs to be actively reinforced through individual learning styles. When students need remediation or corrective instruction, the use of learning styles can increase the chances for individual success. Administrators need to support teachers' instructional efforts and facilitate and encourage their use of a variety of teaching strategies. Different students have different learning styles and needs; therefore, learning styles based instruction is a potentially powerful strategy for increasing mathematics achievement for all students.

REFERENCES

Aiken, Jr., L. R. (1975). Some speculations and findings concerning sex differences in mathematical abilities and attitudes. In E. Fennema (Ed.), *Mathematical learning: What research says about sex differences* (pp. 13–20). Columbus: ERIC Center for Science, Mathematics, and Environmental Education.

Andrews, R. A. (1991, March). *Teach the way they learn.* Paper presented at the meeting of the Association for Supervision and Curriculum Development, San Francisco.

Arkansas Governor's Office. (1992). *A decade committed to change.* Little Rock: Arkansas Department of Education.

Armstrong, J. R. (1975). Factors in intelligence and mathematical ability which may account for differences in mathematics achievement between the sexes. In E. Fennema (Ed.), *Mathematical learning: What research says about sex differences* (pp. 21–31). Columbus: ERIC Center for Science, Mathematics, and Environmental Education.

Armstrong, T. (1987). *In their own way.* Los Angeles: Jeremy P. Tarcher, Inc.

Baratta-Lorton, M. (1976). *Mathematics their way.* Menlo Park: Addison Wesley.

Battista, M. (1981). The interaction between two instructional treatments of algebraic structures and spatial-visualization ability. *Journal of Educational Research, 74,* 337–341.

Behavioral Research Division, Institute for Applied Research Services—University of New Mexico. (1982). *Spatial encounters: Exercises in spatial awareness.* Newton: WEEA Publishing Center.

Benbow, C. P. (1986, April). *Sex-related differences in mathematical reasoning ability among intellectually talented preadolescents: Their characterization, consequences, and possible explanations.* Paper presented at the American Association for the Advancement of Science Conference, Philadelphia.

Benbow, C. P. (1987). Possible biological correlates of precocious mathematical reasoning ability. *Trends in NeuroSciences, 10*(1), 17–20.

Benbow, C. P., & Minor, L. L. (1986). Mathematically talented males and females and achievement in the high school sciences. *American Educational Research Journal, 23,* 425–436.

Benbow, C. P., & Stanley, J. C. (1980). Sex differences in mathematical ability: Fact or artifact? *Science, 210,* 1262–1264.

Berger, R. (1991). Building a school culture of high standards. In V. Perrone (Ed.) *Expanding students assessment* (pp. 32–46). Alexandria: Association for Supervision and Curriculum Development.

Bezusck, S., Kenney, M., & Silvey, L. (1977). *Tessellations: The geometry of patterns.* Palo Alto: Creative Publications.

Black, H., & Black, S. (1981a). *Figural analogies A-1.* Pacific Grove: Midwest Publications.

Black, H., & Black, S. (1981b). *Figural analogies B-1.* Pacific Grove: Midwest Publications.

Black, H., & Black, S. (1981c). *Figural analogies C-1.* Pacific Grove: Midwest Publications.

Black, H., & Black, S. (1983a). *Figural similarities A-1.* Pacific Grove: Midwest Publications.

Black, H., & Black, S. (1983b). *Figural similarities B-1.* Pacific Grove: Midwest Publications.

Black, H., & Black, S. (1983c). *Figural similarities C-1.* Pacific Grove: Midwest Publications.

Black, H., & Black, S. (1983d). *Figural similarities D-1.* Pacific Grove: Midwest Publications.

Blackwell, P. J. (1982). *Spatial encounters: Exercises in spatial awareness.* Newton: Women's Educational Equity Act Publishing Center/EDC.

Borenson, H. (1986). Teaching the process of mathematical investigation. *Arithmetic Teacher,* 36–38.

Brandt, R. (1992). On performance assessment: A conversation with Grant Wiggins. *Educational Leadership, 49* (8), 35–37.

Brandt, R. (1990). On learning styles: A conversation with Pat Guilds. *Educational Leadership, 48* (2), 10–13.

Buerk, D. (1985). The voices of women making meaning in mathematics. *Journal of Education, 167,* 59–70.

Burns, M. (1990). *Mathematics for middle school* [Video]. White Plains: Cuisenaire Company of America.

Burns, M. (1989). *Mathematics with manipulatives* [Video]. White Plains: Cuisenaire Company of America.

Caine, R. N., & Caine, G. (1991). *Making connections: Teaching and the human brain.* Alexandria: Association for Supervision and Curriculum Development.

Campbell, J. (1992). Laser disk portfolios: Total child assessment. *Educational Leadership, 49*(8). 69–70.

Campbell, P. B. (1986). What's a nice girl like you doing in a math class? *Phi Delta Kappan, 67,* 516–520.

Cangelosi, J. S. (1988). Language Activities That Promote Awareness of Mathematics. *Arithmetic Teacher, 36,* 6–9.

Carbo, M. (1992, March). Whole Language and Reading Styles Seminar, Memphis.

Carbo, M., Dunn, R., & Dunn, K. (1986). *Teaching students to read through their individual learning styles.* Englewood Cliffs: Prentice Hall.

Carson, J., & Bostick, R. (1988). *Math instruction using media and modality strengths.* Springfield: Charles C Thomas.

Cherry, C., Goodwin, D., & Staples, J. (1989). *Is the left brain always right?: A guide to whole child development.* Belmont: Fearon Teacher Aids.

Chipman, S., & Wilson, D. (1985). Understanding mathematics course enrollment and mathematics achievement: A synthesis of research. In S. Chipman, L. Brush, & D. Wilson (Eds.) *In Women and Mathematics: Balancing the Equation* (pp. 275–328). Hillsdale: Lawrence Erlbaum.

Chittenden, E. (1991). Authentic assessment, evaluation, and documentation of student performance. In V. Perrone (Ed.) *Expanding student assessment* (pp. 22–31). Alexandria: Association for Supervision and Curriculum Development.

Chow, S. (1992). Using portfolios to assess student performance, *Knowledge Brief* (no. 9). San Francisco: Far West Laboratory.

Claxton, C. S., & Murrell, P. H. (1987). *Learning styles: Implications for improving educational practices* (ASHE–ERIC Higher Education Report No. 4). Washington, DC: Association for the Study of Higher Education.

Dotson, S. (1988, July). *Learning styles in the math classroom.* Paper presented at the eleventh annual leadership institute: Teaching students through their individual learning styles, New York.

Dunn, R. (1988, July). *Teaching students through their individual learning styles: A practical approach.* Paper presented at the eleventh annual leadership institute: Teaching students through their individual learning styles, New York.

Dunn, R., & Dunn, K. (1978). *Teaching students through their individual learning styles.* Englewood Cliffs: Prentice Hall.

Dunn, R., & Dunn, K. (1979). Learning styles/teaching styles: should they . . . can they . . . be matched? *Educational Leadership, 36,* 238–244.

Dunn, R., Dunn, K. & Price, G. E. (1981). *Learning styles inventory.* Lawrence: Price Systems.

Dunn, R., Dunn, K., & Treffinger, D. (1992). *Bringing out the giftedness in your child: Nurturing every child's unique strengths, talents, and potential.* New York: John Wiley & Sons.

Eisele, B. (1991). *Managing the whole language classroom.* Cypress: Creative Teaching Press.

Farr, R. C. (1990, January). *Portfolio Assessment.* Paper presented at the Arkansas Chapter I Conference, Little Rock.

Farr, R. C. (1993). *Portfolio assessment: Teacher's Guide.* Orlando: Harcourt Brace Jovanovich.

Fass, W., & Schumacher, G. M. (1978). Effects of motivation, subject activity, and readability on the retention of prose material. *Journal of Educational Psychology, 70,* 803–807.

Fennema, E. (1975). Mathematics learning and the sexes: A review. *Journal for Research in Mathematics Education, 5,* 126–139.

Fennema, E., & Sherman, J. (1987). Sex-related differences in mathematics achievement, spatial visualization, and affective factors. *American Educational Research Journal, 14,* 51–71.

Finkel, L. G. (1980). *Kaleidoscopic designs and how to create them.* New York: Dover.

Fisher, C., Marliave, R., & Filby, N. (1979). Improving teaching by increasing academic learning time. *Educational Leadership, 39,* 52–54.

Forester, A. D., & Reinhard, M. (1989). *The learners' way.* Winnipeg: Peguis.

Fox, L. H. (1981). *The problem of women and mathematics.* New York: Ford Foundation Report.

Fox, L. H. (1975). Mathematically precocious: Male or female. In E. Fennema (Ed.). *Mathematics learning: What research says about sex differences.* (pp. 1–12). Columbus: ERIC Center for Science, Mathematics, and Environmental Education.

Fox, M. (1993, March). *Using alternative assessment strategies in the classroom.* Paper presented at the Southern Regional National Council of Teachers of Mathematics Conference, Columbus, GA.

Franklin, M. (1990a). *Add-Ventures for girls: Building math confidence. Elementary teacher's guide.* Newton: Women's Educational Equity Act Publishing Center/ECD.

Franklin, M. (1990b). *Add-Ventures for girls: Building math confidence. Junior high teacher's guide.* Newton: Women's Educational Equity Act Publishing Center/ECD.

Frazier, D. M., & Paulson, F. L. (1992). How portfolios motivate the reluctant writer. *Educational Leadership, 49* (8), 62–65.

Frutcher, B. (1954). Measurement of spatial abilities. *Educational and Psychological Measurement, 14,* 387–395.

Grayson, D., & Martin, M. (1988). *Gender/ethnic expectations and student achievement: Teacher handbook.* Earlham: Graymill Foundation.

Gregorc, A. F. (1979). Learning and teaching styles: Potent forces behind them. *Educational Leadership, 36,* 234–236.

Hansen, J. (1992). Literacy portfolios: Helping students know themselves. *Educational Leadership, 49*(8), 66–70.

Harnadek, A. (1979). *Spatial perception – A.* Pacific Grove: Midwest Publications.

Hendrickson, A. D. (1986). Formal reasoning and school mathematics. *School Science and Mathematics, 49*(8), 58–61.

Herbert, E. A. (1992). Portfolios invite reflection – from students and staff. *Educational Leadership, 49*(8), 58–61.

Hetterscheidt, J., Pott, L., Russell, K., & Tchang, J. (1992). Using the computer as a reading portfolio. *Educational Leadership, 49*(8), 73.

Hilton, T. L., & Berglund, G. W. (1974). Sex differences in mathematics achievement – a longitudinal study. *The Journal of Educational Research, 67,* 231–237.

Hunter, M. (1989). Join the "par-aide" in education. *Educational Leadership, 47* (2), 36–37.

Irlen, H. (1991). *Reading through colors.* Los Angeles: Publishers Group West.

Irwin, J. W. (1990). *Teaching reading comprehension processes,* (2nd ed.). Englewood Cliffs: Prentice Hall.

Jarolimek, J. (1991). *Social studies in elementary education.* New York: Macmillian.

Knight, P. (1992). How I use portfolios in mathematics. *Educational Leadership, 49* (8), 71–72.

Kutz, R. E. (1991). *Teaching elementary mathematics.* Boston: Allyn and Bacon.

Lund, C. (1980). *Dot paper geometry with or without a geoboard.* New Rochelle: Cuisenaire Company of America.

Maeroff, G. I. (1991). Assessing alternative assessment. *Phi Delta Kappan, 73,* 272–281.

Maccoby, E. F., & Jacklin, C. N. (1974). *The psychology of sex differences.* Stanford: Stanford University Press, G. I.

Marzano, R. J. (1992). *A different kind of classroom: Teaching with dimensions of learning.* Alexandria: Association for Supervision and Curriculum Development.

McKim, R. H. (1980). *Thinking visually.* Belmont: Lifetime Learning Publications.

Meeker, M. (1979). The relevance of arithmetic testing to teaching arithmetic skills. *The Gifted Child Quarterly, 23,* 297–303.

Midkiff, R. (1991, July). *Math instruction using media and modality strengths.* Paper presented at the Sixth Annual Statewide Mathematics and Leadership Conference, Conway, AR.

Midkiff, R. B., & Towery, R. (1991a, March). *Alternative methods of math and social studies instruction for at-risk students based on learning style needs.* Paper presented at the American Council on Rural Special Education, Nashville.

Midkiff, R. B., & Towery, R. (1991b, November). *Teach them the way they learn.* Paper presented at the Arkansas State Reading Conference, Little Rock.

Midkiff, R. B., Towery, R., & Roark, S. (1991a). Accommodating learning styles needs of academically at-risk students in the library/media center. *Ohio Media Spectrum, 43*(2), 45–41.

Midkiff, R. B., Towery, R., & Roark, S. (1991b). *Learning Style Needs of At-Risk Students: Teaching Math and Social Studies the Way They Learn.* (ERIC Document Reproduction Service No. ED 331 632).

Mitchell, C. E., & Burton, G. M. (1984). Developing spatial ability in young children. *School Science and Mathematics, 84,* 395–405.

Mitchell, S. (1991). *Arkansas math crusade.* Unpublished Manuscript. Arkansas Department of Education, Little Rock, AR.

National Council of Teachers of Mathematics. (1989). *Curriculum and evaluation standards for school mathematics.* Reston: Author.

National Research Council. (1989). *Everybody counts: A report to the nation on the future of mathematics education.* Washington, D.C.: National Academy Press.

O'Neil, J. (1992). Putting performance assessment to the test. *Educational Leadership, 49*(8), 14–19.

O'Neil, J. (1990). Making sense of style. *Educational Leadership, 48*(2), 4–9.

Pasternack, M., & Silvey, L. (1975). *Pattern blocks activities B.* Palo Alto: Creative Publications.

Pattison, P., & Grieve, N. (1984). Do spatial skills contribute to sex differences in different types of mathematical problems? *Journal of Educational Psychology, 76,* 678–689.

Paulson, F. L., Paulson, P. R., & Meyer, C. A. (1991). What makes a portfolio a portfolio? *Educational Leadership, 48*(5), 60–63.

Peterson, P. L., Fennema, E., & Carpenter, T. (1989). Using knowledge of how students think about mathematics. *Educational Leadership, 46*(4), 42–46.

Picciotto, H. (1984). *Pentomino activities lessons and puzzles.* Sunnyvale: Creative Publications.

Research and Educational Planning Center—University of Nevada. (1990). *Add-ventures for girls: Building math confidence—elementary teacher's guide.* Newton, MA: WEEA Publishing Center.

Rindfuss, M. (1993, March). *Using alternative assessment to connect mathematics and other disciplines.* Paper presented at the Southern Regional National Council of Teacher of Mathematics Conference, Columbus, GA.

Rupley, W. H. (1976). Effective reading programs. *The Reading Teacher, 29,* 612–623.

Seymour, D., Silvey, L., & Snider, J. (1974). *Line designs.* Palo Alto: Creative Publications.

Silvey, L., & Taylor, L. (1976). *Paper and scissors polygons.* Palo Alto: Creative Publications.

Stage, E. K., & Kurplus, R. (1981). Mathematical ability: Is sex a factor? *Science, 212,* 114.

Stage, E. K., Kreinbert, N., Eccles, J., & Becker, J. (1985). Increasing the participation and achievement of girls and women in mathematics, science, and engineering. In S. Klein (Ed.), *Handbook for achieving sex equity through education,* (pp. 237–268). Baltimore: Johns Hopkins University Press.

Stangl, J. (1984). *Paper stories.* Belmont: Fearon Teacher Aids.

Stenmark, J. K. (1991). *Mathematics assessment: Myths, models, good questions, and practical suggestions.* Ruston: National Council of Teachers of Mathematics.

Szetela, W., & Nicol, C. (1992). Evaluating problem solving in mathematics. *Educational Leadership, 49*(8), 42–45.

Thomasson, R. D. (1988). The relation of training in spatial perception, arithmetic similarities, logic, and memory of symbolic unit skills to delayed mathematics achievement of female and male gifted students. *Dissertation Abstracts International,* (University Microfilms No. DAO 61522).

Tierney, R. J. (1992, September). Setting a new agenda for assessment. *Learning,* pp. 61–64.

Tobias, S. (1976, September). Math anxiety. *MS Magazine,* pp. 56–59.

Tobias, S. (1978). *Overcoming math anxiety.* New York: W. W. Norton.

Tompert, A. (1990). *Grandfather Tang's story: A tale told with tangrams.* New York: Crown Publishers.

Walker, B. J. (1988). *Diagnostic teaching of reading: Techniques for instruction and assessment.* Columbus: Charles E Merrill.

Webb, G. M. (1983). Left/right brains, teammates in learning. *Exceptional Children, 29,* 508–515.

Weiner, N. C., & Robinson, S. E. (1986). Cognitive abilities, personality and gender differences in math achievement of gifted adolescents. *Gifted Child Quarterly, 30,* 83–87.

Wheeler, C. (1988). Correlation between remedial students and learning styles: Implications for computer assisted instruction. Eastern Washington University, Master's Thesis. (ERIC Document Reproduction Service No. ED 297 294).

Wiggins, M. G. (1992). Creating tests worth taking. *Educational Leadership, 49*(8), 26–31.

Willis, S. D., Wheatley, G. H., & Mitchell, O. R. (1978). Cerebral processing of spatial and verbal-analytical tasks: An EEG study. In M.D. Ainsworth (Ed.) *Neuropsychologia: Vol. 17* (pp. 473–484). New York: Pergamon Press.

Willoughby, S. S. (1990). *Mathematics education for a changing world.* Alexandria: Association for Supervision and Curriculum Development.

Wolf, D. P., LeMahiew, P. G., & Eresh, J. (1992). Good measure: Assessment as a tool for educational reform. *Educational Leadership, 49*(8), 8–13.

Young, J. L. (1981). Improving spatial abilities with geometric activities. *Arithmetic Teacher, 31,* 38–43.

Zessoules, R., & Gardner, H. (1991.). Authentic assessment: Beyond the buzzword and into the classroom. In V. Perrone (Ed.) *Expanding student assessment* (p. 47–71). Alexandria: Association for Supervision and Curriculum Development.

INDEX